Aquatic Habitats in Sustainable Urban Water Management

Science Policy and Practice

T0172822

Urban Water Series - UNESCO-IHP

ISSN 1749-0790

Series Editors

Čedo Maksimović
Department of Civil and Environmental Engineering
Imperial College
London, United Kingdom

Alberto Téjada-Guibert
International Hydrological Programme (IHP)
United Nations Educational, Scientific and Cultural Organization (UNESCO)
Paris, France

Aquatic Habitats in Sustainable Urban Water Management

Science Policy and Practice

Edited by

Iwona Wagner, Jiri Marsalek and Pascal Breil

UNESCO Publishing

Taylor & Francis
Taylor & Francis Group

Cover Illustration

Willowfield stormwater pond, Massey and Taylor Creek System in Toronto, Ontario, Canada, July 19, 2002.
© Quintin Rochfort, Environment Canada. Reproduced with kind permission.

Published jointly by

The United Nations Educational, Scientific and Cultural Organization (UNESCO)
7, place de Fontenoy
75007 Paris, France
www.unesco.org/publishing

and

Taylor & Francis The Netherlands
P.O. Box 447
2300 AK Leiden, The Netherlands
www.taylorandfrancis.com – www.balkema.nl – www.crcpress.com
Taylor & Francis is an imprint of the Taylor & Francis Group, an informa business, London, United Kingdom.

Typeset by Charon Tec Ltd (A Macmillan company), Chennai, India
Printed and bound in Hungary by Uniprint International by (a member of the Giethoorn Media-group).
Székesfehévár.

ISBN UNESCO, paperback: 978-92-3-104062-7
ISBN Taylor & Francis, hardback: 978-0-415-45350-9
ISBN Taylor & Francis, paperback: 978-0-415-45351-6
ISBN Taylor & Francis e-book: 978-0-203-93249-0

Urban Water Series: ISSN 1749-0790

Volume 4

The designations employed and the presentation of material throughout this publication do not imply the expression of any opinion whatsoever on the part of UNESCO or Taylor & Francis concerning the legal status of any country, territory, city or area or of its authorities, or the delimitation of its frontiers or boundaries.
The authors are responsible for the choice and the presentation of the facts contained in this book and for the opinions expressed therein, which are not necessarily those of UNESCO nor those of Taylor & Francis and do not commit the Organization.

British Library Cataloguing in Publication Data
A catalogue record for this book is available from the British Library

Library of Congress Cataloging-in-Publication Data
Wagner, Iwona, 1971–
    Aquatic habitats in sustainable urban water management : science, policy and practice / Iwona Wagner,
Jiri Marsalek, and Pascal Breil.
       p. cm. — (Urban water series, ISSN 1749-0790)
    Includes bibliographical references and index.
    ISBN 978-0-415-45350-9 (hardcover : alk. paper) — ISBN 978-0-415-45351-6 (pbk. : alk. paper) —
ISBN 978-0-203-93249-0 (e-book)
    1. Aquatic habitats—Management. 2. Ecohydrology. 3. Municipal water supply—Management. I. Marsalek, J. (Jiri),
1940– II. Breil, Pascal. III. Title.

QH541.5.W3W34 2008
639.9–dc22

                                                                                            2007032493

Foreword

Aquatic habitats, such as rivers, lakes, ponds and wetlands, provide a range of important ecosystem services and benefits to society. However, the unsustainable use of aquatic habitats, including by the urban water management sector itself, tends to alter and reduce their biodiversity and thus their ability to provide services, including clean water, protection of human health from waterborne diseases and pollutants, protection of urban areas from flooding, and the maintenance of aesthetic and recreational ecosystem services. Spurred by increasing urbanization, population increases and climate change, this is a global issue that is likely to become more serious over the coming years, in particular in the Southern hemisphere. If it is not addressed, there is the threat that several of the Millennium Development Goals will not be reachable.

As a contribution towards addressing this issue, UNESCO, through its International Hydrological Programme (IHP) and its Man and the Biosphere (MAB) Programme, has consequently prepared this book, aiming to improve the understanding of aquatic habitats, their ecosystem goods and services, and their conservation and sustainable use – with a special focus on their integration into urban water management.

This publication, which is part of a series of urban water management books produced in the framework of the Sixth Phase of IHP (2002-2007), is the main output of the project on urban aquatic habitats conducted by IHP and MAB and represents the result of the deliberations of a broad range of experts representing the hydrological, ecological and health sciences. It was skilfully edited by Ms Iwona Wagner, Mr Jiri Marsalek and Mr Pascal Breil and prepared under the responsibility of Mr Peter Dogsé (MAB) and Mr J. Alberto Tejada-Guibert (IHP), who was assisted by the consultant Ms Biljana Radojevic. We are grateful to all the contributors and the editors for their hard work, and we are confident that the conclusions, recommendations and case studies contained in this volume will prove to be of value to urban water management practitioners, policy- and decision-makers and educators alike throughout the world.

András Szöllösi-Nagy
Secretary of UNESCO's International Hydrological Programme (IHP)
Director of UNESCO's Division of Water Sciences
Deputy Assistant Director-General for the Natural Sciences Sector of UNESCO

N. Ishwaran
Secretary of UNESCO's Man and the Biosphere (MAB) Programme
Director of UNESCO's Division of Ecological and Earth Sciences

Contents

List of Figures

Acronyms

ADWR	Arizona Department of Water Resources
AFNOR	Association Française de Normalisation
AHP	Austrian Hydro Power, Inc
AMA	Active Management Area
ASCE	American Society of Civil Engineers
BAHC	Biospheric Aspects of the Hydrological Cycle
BIO	biodiversity
BMPs	Best Management Practices
BOD	Biochemical (or Biological) Oxygen Demand
BR	biosphere reserve
BWR	Basic Water Requirement:
CAP	Central Arizona Project canal
CAPLTER	The Central Arizona – Phoenix Long Term Ecological Research project
CBD	Convention on Biological Diversity
CC	Conservation Councils of Istanbul
CCP	Critical Control Point
CFA	Canada Fisheries Act
CFU	colony forming units
CSOs	Combined Sewer Overflows
CWA	The Clean Water Act
DALY	Disability – Adjusted – Life Year
DCDC	Decision Center for a Desert City
DFO	Department of Fisheries and Oceans
DHKD	The Turkish Society for the Protection of Nature
DO	dissolved oxygen
DWAF	Department of Water Affairs and Forestry
EA	Ecological Ambience
EAP	Environmental Action Plan
EASY	the Ecological Ambiance System
ED	ecosystem defences
EEA	European Environment Agency
EEC	European Economic Community
EHEC	Enterohaemorrhagic *E.coli*
ESI	Environmental Sustainability Index

FAO	Food and Agriculture Organization of the United Nations
FT	Functional Traits
GEP	Good Ecological Potential
GIS	Geographic Information System
GMA	Groundwater Management Act
GV	guideline value
HACCP	hazard analysis and critical control point
IBA	Important Bird Areas of Turkey
IFIM	Instream Flow Incremental Methodology
IGBP	International Geosphere-Biosphere Programme
IHAS	the Integrated Habitat Assessment System
IJC	International Joint Commission
	organic and mineral inputs
IOBS	indice oligochètes de bioindication des sédiments (oligochaete index for sediment biomonitoring)
IPA	Important Plant Areas of Turkey
ISKI	Istanbul Water and Sewage Board
IUWM	Integrated Urban Water Management
IWA	International Water Association
IWRM	integrated water resource management
LID	Low Impact Development
LOUE	Lowest Observed Urbanisation Effect
MEA	Millennium Ecosystem Assessment
MAB	Man and the Biosphere Programme of UNESCO
MCA	Multi Criteria Analysis
MDG	Millennium Development Goals
MEP	Maximum Ecological Potential
	Municipality of Greater Istanbul
	Ministry of Environment and Forestry -Nature Protection Directory
NEPA	The National Environmental Protection Act
NPDES	The National Pollutant Discharge Elimination System
	The Omerli Watershed
	Public Awareness/Education/Participation Programmes
PCB	polychlorinated biphenyls
PEC	predicted environmental concentration
PPCP	pharmaceuticals and personal care products
RDM	Resource Directed Measures
RHS	River Habitat Survey
RIVPACS	River Invertebrate Prediction and Classification System
SERCON	System for Evaluating Rivers for Conservation
SRF	short rotation forestry
SRP	Salt River Project
SS	suspended solids
SUDS	Sustainable Urban Drainage Systems
SWITCH	Sustainable Water Management Improves Tomorrow's Cities' Health

TMUWC	Total management of the urban water cycle
UAE	United Arab Emirates
UAHM	Urban aquatic habitat management
UASB	Upflow Anaerobic Sludge Blanket
UBR	Urban Biosphere Reserve
UAH	Urban Aquatic Habitat
UNCHS	United Nations Center for Human Settlements (HABITAT)
UNDSD	United Nations Division for Sustainable Development
UNEP	United Nations Environment Programme
UNEP-GPA	United Nations Environment Programme, Global Programme of Action
UNESCO-IHE	United Nations Educational Scientific and Cultural Organization, Institute for Water Education
USACE	US Army Corps of Engineers
USEPA	US Environmental Protection Agency
WFD	Water Framework Directive
WPCF	Water Pollution Control Federation
WSP	Water Safety Plant
WSUD	Water Sensitive Urban Design
WWTP	Wastewater Treatment Plant

Disability Adjusted Life Years (DALY)   a health gap measure that extends the concept of potential years of life lost due to premature death to include equivalent years of healthy life lost by virtue of being in state of poor health or disability. (*Source*: WHO)

ecohydrology   a sub-discipline of hydrology that focuses on ecological processes occurring within the hydrological cycle and strives to utilize such processes for enhancing environmental sustainability.

ecological potential   the status of a heavily modified or artificial water body measured against the maximum ecological quality it could achieve. It is intended to describe the best approximation to a natural aquatic ecosystem that could be achieved given the hydromorphological characteristics that cannot be changed without significant adverse effects on the specified use or the wider environment.

ecotone   narrow and fairly sharply defined transition zone between two or more ecosystems, which is typically species rich.

eutrophication   enrichment of water by nutrients, especially compounds of nitrogen and phosphorus, that increases productivity of ecosystems, leading usually to lowering water quality and several adverse ecological and social effects (e.g., secondary pollution due to accelerated growth of algae and toxic cyanobcateria, depletion of oxygen).

gabion   wire basket, filled with stones, used to stabilize banks of a water course and to enhance habitat.

Geographical Information System (GIS)   a computer-based system of principles, methods, instruments and geo-referenced data used to capture, store, extract, measure, transform, analyse and map phenomena and processes in a given geographic area.

habitat   the place or type of site where an organism or population naturally occurs.

low impact development   a landscape management approach used to replicate or restore natural watershed functions and/or address targeted watershed goals and objectives.

phytoremediation   the use of plants to remove or deactivate contaminants or pollutants from either soils (e.g., polluted fields) or water resources (e.g., polluted lakes).

polychlorinated biphenyls (PCB)   PCBs are mixtures of more than 200 different organochlorine chemical substances (cogeners) that are known for toxicity and persistence.

preservation   the protection of the natural environment from unnatural disturbance.

rehabilitation   the return of a degraded ecosystem to a non-degraded condition but which may also be different from its original condition.

remediation   cleanup or other methods used to remove a toxic spill or hazardous materials from a site.

reno mattress   a shallow, wide, flexible woven-wire basket (one type of gabion) composed of two to six rectangular cells filled with small stones (often used at culvert inlets and outlets to dissipate energy and prevent channel erosion).

restoration   a management process striving to re-establish the structure and function of ecosystems as closely as possible to the pre-disturbance conditions and functions.

riffle   shallow rapids in an open channel, where the water surface is broken into waves by obstructions totally or partly submerged.

short rotation forestry   growing trees as a crop, including site preparation and management practices used in agriculture, for obtaining high biomass in a short time, to be used as energy crop for use in power stations.

soft approach to water management   a comprehensive approach to water management relying on problem prevention rather than "hard" structural measures, and involving ecosystem biotechnologies.

urban drainage   a system of conveyance and storage elements serving to drain urban areas.

wastewater management   collection, treatment (chemical, biological and mechanical), and reuse of municipal and industrial wastewaters serving to remove and/or reduce their pollution prior to discharge into receiving waters.

Water Framework Directive (EU)   Directive 2000/60/EC establishing a framework for the European Community action in the field of water policy. It aims to secure the ecological, quantitative and qualitative functions of water. It requires that all impacts on water will have to be analysed and actions will have to be taken within river basin management plans.

water quality   physical, chemical, biological and organoleptic (taste-related) properties of water.

water reuse   the process of using reclaimed water or wastewater for beneficial purposes, such as agricultural and landscape irrigation, and industrial processes.

Chapter 1

Introduction to urban aquatic habitats management

Pascal BREIL[1], Jiri MARSALEK[2], Iwona WAGNER[3], Peter DOGSE[4]

[1] Cemagref, Biology (1) and Hydrology/Hydraulics (3) Research Units, 3bis quai Chauveau, C.P 220, F-69336 Lyon Cedex 09 France
[2] National Water Research Institute, Environment Canada, Burlington ON, Canada L7R 4A6
[3] European Regional Centre for Ecohydrology under the auspices of UNESCO, Polish Academy of Sciences, 3 Tylna Str, 90-364 Lodz, Poland Department of Applied Ecology University of Lodz, 12-16 Banacha Str, 90-237 Lodz, Poland
[4] UNESCO Man and Biosphere Programme, 1, rue Miollis, 75732 Paris Cedex 15, France

1.1 IMPACT OF GLOBAL PROCESSES ON WATER RESOURCES IN CITIES

Urban populations are growing worldwide: more than 54% of the world's population of about 6.5 billion people currently live in urban areas. Furthermore, the distribution of urban and rural populations is not uniform – in some countries, more than 90% of residents live in cities. This trend is particularly pronounced in the developing world, where 15 of the world's 20 megacities (cities with more than 10 million inhabitants) are located, and will continue to grow rapidly during the next few decades (Marshall, 2005).

The United Nations (UN) estimates that, at present, the world's urban population grows at a rate of around 180,000 every day. This means that the world's urban infrastructure and water resources, including urban aquatic habitats (UAH), have to absorb the impact generated by the equivalent of double the population of Tokyo – the world's largest city, with 35 million inhabitants – each year. According to the International Water Association (2002), this is a consequence of a combination of intensive human population growth, rural-to-urban migration on a scale exceptional in the history of civilisation, and the annexation of rural areas by cities, and is one of the major causes of the world's water crisis.

According to the UN's *World Urbanization Prospects* (2003), this process will continue in the future. While mega-cities like Tokyo, Mexico City and New York will still dominate the urban landscape in some countries, it is the smaller urban settlements (with fewer than 500,000 inhabitants) in the less developed regions of the world that will be absorbing most of the intensive growth impact. Moreover, the majority of urban dwellers will reside in smaller cities. This means that much of the action on urban water infrastructure and resources, including new approaches to urban aquatic habitats management (UAHM), will have to be taken in mid-size cities (Cohen, 2006).

Urban populations cause large demands on life-support resources and services, including water. The provision of life necessities is becoming a major challenge in all parts of the world. It is well recognized that many resources can be provided by sustainable aquatic habitats, which have the potential to produce and sustain a range of ecosystem services of great importance for economic development and human welfare. These include renewable supplies of fresh water, waste treatment by self-purification in receiving waters, land irrigation, local climate regulation, the buffering of some climate change impacts, educational and recreational services, biodiversity maintenance, and the provision of food, fuel and fibre, to name but a few. However, a broad range of direct and underlying effects of increasing urban pressures may threaten the ability of aquatic habitats to provide such support (Millennium Ecosystem Assessment, 2005). These issues have been widely documented and discussed in many publications, including the book on Urban Water Cycle Processes and Interactions of the UNESCO series on Urban Water Management (Marsalek et al. 2007), which discusses the physical, chemical, biological and combined impacts of urbanization on all environmental fields, including the atmosphere, surface waters, wetlands, soils, groundwater and biota.

In spite of the adverse impacts of urbanization and the challenges associated with living in densely populated urban areas, there are many benefits to urban life that attract rural population. Such benefits include opportunities for social and economic development, modern styles of living with high female labour force participation, relatively high levels of indicators of general health, well-being and literacy, and a small ecological footprint (Cohen, 2006). At the same time, many disadvantages and negative impacts may be efficiently mitigated by sustainable management strategies. Next to population growth and the associated land-use transformations, climate change is the second biggest challenge in current water management. It is well documented in the most recent Intergovernmental Panel on Climate Change (IPCC) report (IPCC, 2007) that the global climate is changing, with increasing temperatures and many other associated consequences. Generally speaking, these consequences include modifications of precipitation patterns and more frequent weather extremes, leading to more frequent and less predictable floods and droughts. Still, specific predictions, particularly at small, regional scales, contain large inherent uncertainties (IPCC, 2007). A combination of multiple stressors contributes to the degradation of the environment and lowers the predictability of environmental processes, including those taking place in aquatic habitats.

1.2 AQUATIC HABITATS IN INTEGRATED URBAN WATER MANAGEMENT: HOW ARE THEY MANAGED OR MISMANAGED?

Aquatic habitats are defined here as water bodies supporting aquatic life. Increased effluent temperatures, greater discharges of water, pollutants and waste, and changes in water bodies' morphology have an impact on all of the basic characteristics of habitats and affect the performance of associated biological communities. Affected ecosystems lose their resistance to escalating stresses as well as their ability to adapt to changing conditions. Perhaps the most important among the basic habitat characteristics is the flow regime – or, stated differently, the availability of water in adequate time, space and quantity. Indeed, aquatic habitats obviously need water to function, and unless their water requirements are met, their capacity to produce and sustain ecosystem services is seriously impaired.

Water stress imposes losses of, and trade-offs between, services and their different beneficiaries and stakeholders. Considering growing levels of water use and wastewater production, the need for tradeoffs between meeting the demands of urban populations and serving the needs of aquatic habitats will be even more challenging in the future. The dilemmas faced by countries with acute and increasing water shortages, further exacerbated by growing populations and rising water consumption, are particularly daunting and extremely complex (Maksimovic and Tejada-Guibert, 2001). A balanced approach that addresses these pressing issues is not common in the water service sector, which has been traditionally based on a sectoral approach. Conventional urban water management has focused mainly on protecting the urban human population against hydrological extremes (floods and droughts) and providing water services. It has come to include water supply, urban drainage and flood protection, wastewater management and, more recently, some form of aquatic ecosystems protection. This, however, often does not address specific features of aquatic habitats, their needs and potentials. In many cases, to minimize drainage costs, urban streams and rivers have been incorporated into major drainage systems and have conveyed various types of municipal effluents, resulting in extreme habitat degradation.

Currently, the development of comprehensive knowledge generated by the integration of various sectors of science as well as recent developments in ecological engineering, are increasing opportunities to develop more sustainable, economically viable urban environments. Newly emerging paradigms underline the need for water conservation, rational use, reuse and the sustainable integration of different components of urban river systems, including those of a technical and natural character (Pinkham, 2004; Zalewski, 2006). This tendency creates opportunities for changing attitudes to UAHs, and their use for concurrently improving efficiency of urban water management and the quality of human life in cities (Zalewski and Wagner, 2006).

The approach taken in this book examines specific features of aquatic habitats in the urban environment and the means of their protection and sustainable exploitation. In that sense, it fits well into general environmental strategies, including sustainable development and its implementation using ecosensitive measures and ecosystems approaches. It is further recognized that the change in attitude towards better protection of urban aquatic habitats needs to involve all stakeholders, including the general public (often organized in citizen groups or NGOs), water management professionals and scientists, as well as decision makers and policy makers at all levels of government. However, this publication does not deal with economic, socio-economic and institutional issues specific to urban water management. Those matters were introduced earlier in a UNESCO publication by Maksimovic and Tejada-Guibert (2001) and are covered in this series of the UNESCO publications on Integrated Urban Water Management (IUWM). This book is intended to draw the attention of urban water stakeholders to the importance of aquatic habitats in general, while presenting concepts, approaches, technical solutions and research needs, integrating the broad range of ecosystem services derived from aquatic habitats.

This book focuses on urban aquatic habitats, including streams, rivers, wetlands, ponds, reservoirs, impoundments and lakes. These habitats interact with other components of the urban or peri-urban water cycle and infrastructure and should therefore be properly considered in IUWM. Water input into these habitats occurs by precipitation, import of water (usually inflow from upstream watersheds, or by pumping and conveyance systems), withdrawal of groundwater, and discharges from such sources as stormwater runoff, combined sewer overflows (CSOs) or both treated and untreated wastewater effluents. The flow exchanges between various components, such as surface water and groundwater or sewer pipe seepage, are often underestimated. Anthropogenically modified flows affect aquatic habitats' hydrological regime, physical structure and water quality. Pollution can have effects on both humans and water biota that are instantaneous (acute toxicity), chronic (manifest over a longer time) or cumulative (resulting from various types of impacts and their persistent cumulative action in time). Such processes, and many others, can be, to a certain degree, controlled and reduced by understanding UAHs' functioning, strengthening their self-defence mechanisms, integrating them into the urban management practices and synchronizing their use with technical measures (e.g., urban infrastructure exploitation).

1.3 COMMENTS ON URBAN AQUATIC HABITAT MANAGEMENT: RESTORATION, PRESERVATION, REHABILITATION OR REMEDIATION?

The severe degradation of urban aquatic habitats creates constraints for their management, such as those resulting from dense development of adjacent lands or high urban property prices. Whatever realistic management objectives for UAHs are agreed on, the decision has to result from considering several criteria, such as land-use, natural resources planning, peoples' perception of the existing and desirable states of the urban environment, availability of resources and local policy goals. Hulse and Gregory (2004) describe the decision-making process as looking for a balance between two extremes of the ecosystem state – an ecosystem having low ecological potential and being burdened with high ecological and economic constraints on one side, and an ecosystem with high ecological potential and low demographic and economic constraints on the other. Criteria for making a decision should balance a potential increase in ecological benefits (and possibly of human well-being) and spatial, demographic and economic limitations, together with economic gains and losses.

Among several approaches to urban aquatic habitats, the following are usually considered:

- **Restoration** is a process that, ideally, brings a degraded river back to its original conditions. It includes restoring water quality, sediment and flow regime, channel morphology, communities of native aquatic plants and animals, and adjacent riparian lands. The goal of restoration is impossible to achieve in urban watercourses. Re-establishing the habitat's historical, original state, however, would require the replication of original conditions, which no longer exist, and are not even well known.
- **Preservation** of aquatic habitats state and biodiversity is a realistic goal, when the urbanization impact on ecosystems is not severe. This ideal situation generally occurs in peri-urban areas, where urbanization has not yet fully invaded the surrounding landscapes, and where industrial or agricultural activities are limited (Lafont et al., 2006).
- **Rehabilitation** is a less ambitious but more realistic aim. It enhances or re-establishes lost or diminished biotic functions of ecosystems that can persist without attempts to restore pristine conditions. It improves the most important aspects of aquatic environment and creates habitats resembling original conditions.

- **Remediation** is an approach applied in cases where environmental changes are irreversible and catchment conditions no longer support aquatic ecosystem functioning. Remediation aims to improve ecological conditions in the aquatic ecosystem, which may not lead to a state resembling the original state of the stream. It means that after the remediation process, we can obtain a new ecosystem, different from the original one (Lovett and Edgar, 2002).

1.4 STRUCTURE OF THIS BOOK

This book is meant to introduce water management professionals to often neglected aspects of integrated urban water management, including the protection of urban aquatic habitats and, where feasible, the utilization of aquatic habitat properties in the delivery of urban water services. For this purpose, it focuses on presenting both general and specific principles used in UAH management and provides numerous references to other sources of detailed information on this topic. It was prepared by an international team of authors with broad experience in the field from various regions of the world, and the authors were further guided by UNESCO staff, ensuring coverage of the aspects related to both developed and developing countries.

The book can be divided into three parts. The first, part (Chapters 1 to 3) deals with the review of basic concepts and challenges encountered in UAHs, as well as general strategies for integrating aquatic habitats into urban water management. The second part (Chapters 4 to 8) provides the reader with technical measures related to urban water habitats management and rehabilitation, as well as their incorporation into urban planning and their role in human health. Finally, the third part of the book (Chapter 9) gives practical examples of existing UAH issues and possible approaches to solving them, presented in the form of case studies from all over the world.

Following the Introduction to Urban Aquatic Habitats Management, Chapter 2 presents a description of urban aquatic habitats and problems and challenges faced in their management in urban areas. It presents potential benefits from integration of the UAH into the management of water resources in cities. Towards this end, a conceptual model linking the magnitude of human impacts on and the biodiversity of aquatic ecosystems is proposed. It helps to visualize the functioning of the temporal dynamics of bioassimilation processes that need to be considered in urban water management. Such an approach helps to set realistic objectives for management actions, and differentiate among various approaches in UAHM presented in the previous section. Setting realistic management actions has to be based on the state-of-the-art strategies as well as relevant policies and regulations. These aspects are presented in Chapter 3: Strategies, Policies and Regulations Integrating Protection and Rehabilitation of Aquatic Habitats in Urban Water Management. The necessity of addressing urban environmental problems in an integrated manner has led to the development of several concepts, including sustainable development, ecosystem approaches and performance criteria and indicators for major pillars (society, environment, economy and technical realization) considered in the implementation of sustainable solutions. Three specific management approaches applicable to urban environment components are discussed in some detail: low impact development, total management of the urban water cycle and soft path for water. These concepts are then supplemented by examples of advanced policies and regulations promoting the health of aquatic habitats, such as the European Union's Water Framework Directive.

The second part of the book starts with a presentation of management strategies and is followed by technical measures for ecosensitive management of urban aquatic habitats in Chapter 4: Ecosensitive Approaches to Managing Urban Aquatic Habitats and their Integration with Urban Infrastructure. Specific technical measures addressed include those concerning interactions of UAHs with water supply systems (including the issues of water import and river damming), urban drainage and flood protection (including stormwater management and flood protection), and wastewater management and sanitation (including wastewater collection, treatment and reuse). The chapter provides examples of ecosensitive approaches meeting multiple, sometimes conflicting, goals, where aquatic habitats may serve the urban population needs without losing their own ecological values and functions.

The highest environmental and societal benefits of integration of aquatic habitats with the city infrastructure can be obtained while maintaining UAHs at their highest ecological potential. In many circumstances, the state of aquatic habitats is so degraded that they cannot be rehabilitated by natural processes alone. In such cases, technical intervention, including reconstructing the physical habitat structure, is necessary. The importance of the issues of habitat rehabilitation or restoration are addressed separately, and in detail, in Chapter 5: Aquatic Habitat Rehabilitation: Goals, Constraints and Techniques, which features methods for the assessment of the ecological potential of a river, and techniques for restoration of its hydrological characteristics, and physical and biotic structures. Finally, it proposes four stages of an adaptive approach to river restoration, which greatly increases the likelihood of success for restoration projects.

Urban aquatic ecosystems are usually exposed to highly concentrated impacts, while degradation of their physical structure, and thus ecological functions, lowers their capacity to cope with the stress. Enhancing the capacity of aquatic ecosystems to absorb the impacts, while keeping the costs of management low, can be achieved by ecohydrology, as discussed in Chapter 6: Ecohydrology of Urban Aquatic Ecosystems for Healthy Cities. Ecohydrology postulates the use of dual regulation between hydrological and biota dynamics as a management tool. Synergistic integration of ecohydrological solutions into the urban catchments and their harmonization with engineering solutions may lower the overall costs of management and improve quality of life and human health.

Chapter 7: Integrating Aquatic Habitat Management into Urban Planning deals with the integration of well-functioning aquatic habitats into urban landscape planning. In spatial planning, strategically placing flowing and still waters, defined as rivers and wetlands, is of utmost importance and relies primarily on a functional, spatially-oriented inventory of the aquatic habitats in the urban landscape. One key issue is the preservation or rehabilitation of the hydrological corridor in an urbanized environment. Arguments for ecosystem integrity are also presented through goods and services that UAHs can provide in an urban context. Buffer and set-back zone design, including their location, management and benefits, provide concrete ideas on this issue. The importance of social and economic factors in urban ecology development and maintenance is emphasized.

One of the fundamental arguments for developing a new management approach and reversing UAH degradation in cities is the potential impacts on human population. Those issues, with respect to human health, are addressed in Chapter 8: Human health and safety related to urban aquatic habitats. Human health in cities has become one of the major issues arising from rapid ongoing urbanization, as highlighted by the UN Millennium Development Goals. Considering the importance of this issue for

UNESCO, a separate chapter is devoted to water-related health risks and their mitigation in cities. Well-managed aquatic habitats can create positive feedback improving health and quality of life in cities. This chapter also provides basic information about the development of Water Safety Plans (WSP) using the Hazard Analysis and Critical Control Point (HACCP) approach.

Finally selected case studies from a number of cities around the world are presented in Chapter 9. These case studies represent a broad range of situations found in urban areas, including a rehabilitation and flood management project for a small river in Cape Town, South Africa; the rehabilitation of a salty marsh reserve in Sharjah, United Arab Emirates; an ecohydrological approach to rehabilitating the City of Lodz, Poland; the integration of ecological and hydrological issues into urban planning in the Adige River corridor in Italy; the assessment of urban stream bio-assimilation capacity to cope with combined sewer overflows in Lyon, France; the maintenance of floodplain wetland biodiversity in Vienna, Austria; a proposal to designate an urban catchment as a Urban Biosphere Reserve in Istanbul, Turkey; and an overview of ecology and water management issues in the metropolitan area of Phoenix, USA. Overall, these case studies provide examples of UAH issues with respect to developing solutions, planning and implementation processes, involving stakeholders, the selection of preferred solutions, and for the completed cases, an assessment of the final product. They provide guidance for addressing other similar cases elsewhere and demonstrate the benefits of the planning and technical measures discussed in the book.

In conclusion, numerous benefits of well-managed urban aquatic habitats are undisputable and should inspire urban water managers to include urban aquatic habitat protection and habitats' beneficial utilization in their efforts to achieve a better quality of life in cities and healthier aquatic ecosystems. It is up to the urban water stakeholders to seize these benefits and opportunities by incorporating UAHs into urban water management with the aim of making the habitats and life in cities more sustainable.

REFERENCES

Cohen, B. (2006). Urbanization in developing countries: current trends, future projections, and key challenges for sustainability. *Technology in Society*, 28, 63–80.

Hulse, D. and Gregory, S. 2004 Integrating resilience into floodplain restoration. *Journal of Urban Ecology*, Special Issue on Large-Scale Ecosystem Studies: Emerging trends in urban and regional ecology. Vol. 7, No. 3–4.

IPCC (Intergovernmental Panel on Climate Change) 2007. *Climate Change 2007: The Physical Science Basis. Summary for Policymakers*. IPCC Secretariat, WMO, Geneva, Switzerland.

IWA (International Water Association). 2002.

IPCC. 2007. *Climate Change 2007. 4th Assessment Report*. Intergovernmental Panel on Climate Change. WMO, UNEP.

Lafont, M., Vivier, A., Nogueira, S., Namour, P. and Breil, P. 2006. Surface and hyporheic Oligochaete assemblages in a French suburban stream. *Hydrobiologia* 564: 183–93.

Lovett, S. and Edgar, B. 2002. 'Planning for river restoration', Fact Sheet 9, Land & Water Australia, Canberra.

Maksimovic, C. and Tejada-Guibert, J.A. 2001. Frontiers on Urban Water Management: Deadlock or Hope? IWA Publishing, ISBN: 1 900222 76 0, London, UK.

Marsalek, J., Jimenez-Cisneros, B., Karamouz, M., Malmqvist, P.A., Goldenfum, J. and Chocat, B. (2007). *Urban water cycle processes and interaction*. Taylor & Francis Group, London, UK.

Marshall, J. (2005). Megacity, mega mess ... *Nature*. Sept. 15, 2005, 312–314.

Millennium Ecosystem Assessment, 2005. Ecosystems and Human Well-being: Wetlands and Water Synthesis. World Resources Institute, Washington DC.

United Nations. 2003. *World Urbanisation Prospect. The 2003 Revision.*

Pinkham, R., 2004. 21st Century Water Systems: Scenarios, Visions, and Drivers. http:// www. rmi.org/images/other/Water/W99-21_21CentWaterSys.pdf. Rocky Mountain Institute, Snowmass, Colorado

Zalewski, M. 2006. Ecohydrology – an interdisciplinary tool for integrated protection and management of water bodies. Arch. Hydrobiol. Suppl. 158/4, pp. 613–22.

Zalewski M. and Wagner I. 2006. Ecohydrology – the use of water and ecosystem processes for healthy urban environments. Aquatic Habitats in Integrated Urban Water Management. *Ecohydrology & Hydrobiology*. Vol. 5. No 4, pp. 26–268.

Chapter 2

Urban aquatic habitats: Characteristics and functioning

Michael LAFONT[1], Jiri MARSALEK[2] and Pascal BREIL[1]

Cemagref Biology and Hydrology, Hydroecology Research Unit, 3bis quai Chauveau CP 220
F-69336 Lyon Cedex 09 France

National Water Research Institute Environment Canada, 867 Lakeshore Rd, Burlington Ontario L7R
4A6 Canada

The need for protecting or rehabilitating aquatic habitats in urban areas and the need for incorporating such management actions into urban water management have been briefly introduced in Chapter 1. The main purpose of this chapter is to further develop the urban aquatic habitats description, with respect to their basic characteristics and functioning, and set the stage for further discussions of their management in the following chapters. Understanding the habitat characteristics that affect aquatic biological communities, as well as the functioning of the aquatic ecosystem, helps to set realistic objectives for the protection or rehabilitation of its ecological status or potential and set implications for management actions. Various aspects of aquatic habitat protection and rehabilitation and their integration with urban water management with respect to strategies, planning and technical measures are addressed later on in this book.

2.1 CHARACTERISTICS OF AQUATIC HABITATS

The issues surrounding aquatic habitats and the life they support are very broad and have been somewhat reduced for the purpose of this book by focusing on freshwater (i.e. water bodies fully surrounded by urban land, as opposed to coastal regions). Urban aquatic habitats are defined here as natural or constructed freshwater bodies, defined by their physical features, and, as mentioned in the previous chapter, they include urban streams, canals, rivers, ponds, impoundments, reservoirs and lakes. In some later chapters, the discussion is expanded to include some related aspects of ecotone habitats.

Habitats, defining physical boundaries of aquatic ecosystems, create conditions for the development and functioning of their biotic component, aquatic life. Freshwater aquatic life includes all species of plants, animals and micro-organisms that, at some stage of their lifecycles, must live in freshwater. Among the various forms of aquatic life, much of the discussion focuses on fish, benthic invertebrates and phytoplankton and, to some extent, on vascular flora typically found in riparian ecotones of freshwater bodies.

The main aquatic habitat characteristics fall into five groups: flow regime, physical habitat structure, chemical variables (water quality), energy (food) sources and biotic interactions, (see Figure 2.1).

Such characteristics are similar to those listed, in much greater detail, in the EU Water Framework Directive (Directive 2000/60/EC; see Chapter 3 for more details).

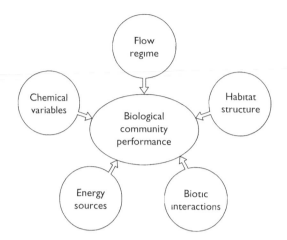

Factors influencing biological community performance
Source: Yoder, 1989.

The EU Water Framework Directive (EU WFD) lists the following variables describing the quality of river and lake ecosystems:

R ...

* Biological elements (composition and abundance of aquatic flora and benthic invertebrate fauna; composition, abundance and age structure of fish fauna)
* Hydromorphological elements supporting the biological elements (hydrological regime; quantity and dynamics of water flow; connection to groundwater bodies; river continuity; morphological conditions; river depth and width variation; structure and substrate of the river bed; and structure of the riparian zone)
* Chemical and physico-chemical elements supporting the biological elements (general; thermal conditions; oxygenation conditions; salinity; acidification status; nutrient conditions), and
* Specific pollutants (pollution by all priority substances identified as being discharged into the body of water; pollution by other substances identified as being discharged in significant quantities into the body of water).

Lak -

* Biological elements (composition, abundance and biomass of phytoplankton; composition and abundance of other aquatic flora; composition and abundance of benthic invertebrate fauna; composition, abundance and age structure of fish fauna)
* Hydromorphological elements supporting the biological elements (hydrological regime; quantity and dynamics of water flow; residence time; connection to the ground water body; morphological conditions; lake depth variation; quantity, structure and substrate of the lake bed; structure of the lake shore)
* Chemical and physico-chemical elements supporting the biological elements (general transparency, thermal conditions, oxygenation conditions, salinity, acidification status, nutrient conditions; specific pollutants)

Physical features of aquatic habitats being affected by human pressures in urban catchments

Flow regime	Physical habitat structure	Chemical variables water quality	Food/energy sources	Biota biotic interactions
• water depth • flow velocity • flow variability • discharge • flood magnitude • flood frequency • drought frequency	• habitat diversity • habitat connectivity • channel sinuosity • siltation • sedimentation pattern • bank stability • substrate type • plant cover	• nutrients • thermal regime • toxins • pseudo-hormones • salinity • turbidity • oxygen concentration • pH	• primary production of algae and macrophytes • energy • particulate organic matter • aquatic and terrestrial invertebrates	• species structure (invasions of exotics) • endemic species • threatened, endangered, sensitive species • species richness • trophic structure • age and genetic structure • predation • competition

Source: Lapińska, 2004; Karr et al., 1986, modified.

* Specific pollutants (pollution by all priority substances identified as being discharged into the body of water; pollution by other substances identified as being discharged in significant quantities into the body of water).

The assessment of the physical quality of habitats is based on the quality and degree of change of the above factors shown in Figure 2.1 and listed in the EU WFD, usually in comparison to reference (natural/un-impacted) conditions. It should be also understood that any modification of the physical characteristics of a habitat is followed by changes in the aquatic biota's structure and performance. Therefore a proper assessment of freshwater ecosystems quality goes beyond physical characteristics. It should also include the assessment of the biotic compartment, which is usually based on relevant biotic indices (see Chapter 5 for a more detailed discussion).

2.2 HUMAN MODIFICATIONS OF AQUATIC HABITATS

The characteristics of aquatic habitats presented in the above section are also reflected in the classification of major human pressures to which they are exposed (see Table 2.1). Aquatic habitats in urban areas are particularly exposed to these pressures. The primary factors of habitat deterioration are water quality decline and flow alternations, which are particularly pronounced in urban catchments. The broad range of chemicals from point and diffused sources include toxic substances and carcinogens, fertilizers, pesticides, herbicides, household hazardous wastes, oils, anti-freeze, heavy metals, pet and yard wastes, and pseudo-hormones (see Chapter 8 for a more detailed discussion of these issues and the related potential risks for urban populations). Simultaneously, the natural heterogeneity of habitats is often degraded or destroyed. In many cities, aquatic habitats are included in the urban drainage system, receiving and disposing of stormwater. The structure of aquatic habitats is adapted for these purposes by such measures as canalization, lining with concrete slabs, water intakes, and damming, among others. These modifications induce the impacts described in the later sections, as well as changes in habitat characteristics.

2.2.1 Flow regime

There is a general agreement that flow regime controls, to a great extent, affects the development of the physical habitat structure in non-channelized rivers, with respect to both macro- and micro-habitat features, while creating variable conditions for the associated aquatic life. Analysis of flow regimes should focus on the aspects that are most important for creating and maintaining the habitat's physical structure and supporting the requirements of aquatic organisms during various life stages. High flows are important for stream geomorph-ology, bed erosion and sediment transport; affect fish migration; and create periodical connectivity between the river and its floodplains, thereby providing additional spawning grounds for biota and juvenile rearing (Horner et al., 1994; Poff et al., 1997). Low flows usually have a greater impact on water quality, for example by lowering dissolved oxygen (DO), increasing temperature or increasing concentrations of many pollutants (Spence et al., 1996). Moreover, in the case of long-lasting and deep droughts, low flows may also set limits on the proper functioning of the ecological processes in river ecosystems.

Urbanization greatly affects the hydrological cycle and by extension the flow regime as well. Such changes are summarized by Roesner et al., (2005) as more frequent and higher flows, increased duration of geomorphically significant flows (MacRae, 1997), flashier/less predictable flows (Henshaw and Booth, 2000), altered timing and rate of change relative to riparian and floodplain connections (Poff et al., 1997), altered duration of low-flow periods (Finkenbine et al., 2000), and conversion of subsurface distributed discharge (interflow) to surface point discharge (Booth and Henshaw, 2001).

The management of urban habitats usually implies high levels of flow control, particularly intermediate and large-scale flood flows, and river channelization. This limits the active role of flow regime in the creation of habitat diversity. Modification of the flow regime causes changes in the sediment transport and stream geomorphology. The isolation of surface water bodies from ground waters and floodplains reduces water storage capacity and results in dramatic changes of discharge patterns. This increases freshwater ecosystems' susceptibility to droughts, and short and extremely high peak flows impact their stability. Frequently, magnitude and irregularity of events change environmental templates, creating conditions unsuitable for the inhabiting organisms. Intended or unintended modifications of flow regime should be examined for key features of the creation and maintenance of the habitat's physical structure, supporting the requirements of aquatic organisms during various life stages.

In some climates, there are also intermittent streams (often described by such terms as arroyos, wadis or torrents), which flow only during a part of the year. Nevertheless these streams are also ecologically important and can be characterized by special biological assemblages (algae, bryophytes, invertebrates and amphibians) that are indicative of the degree of hydrologic permanence. Finally, there are also ephemeral streams, which form only during and after rainfall. A similar classification could be used for impoundments with intermittent water storage.

2.2.2 Physical habitat structure

The physical habitat structure in water bodies is important for the structure and functioning of aquatic communities. The physical habitat structure in water bodies inhabited by aquatic organisms results from interactions between natural geomorphology, flow and waves, sediments entering the water body, and riparian vegetation, and is characterized by macro- and micro-habitat features (Bain and Stevenson, 1999). Macro-habitats, also referred to as hydraulic biotopes (Wadeson and Rowntree, 1998) are classified into fast- and slow-water macro-habitats; the former containing features such as low-gradient stream sections or riffles, and high-gradient sections with rapids, cascades, falls, steps, or chutes, and the latter including pools, straight scours, backwater eddies, plunge pools, and dammed or abandoned channels (Bain and Stevenson, 1999). Micro-habitats are usually described by substrate type, cover, depth, hydraulic complexity and current velocity (Spence et al., 1996). Substrate is generally classified according to the particle size, ranging from silt and clay (<0.059 mm), to sand (0.01 to 1 mm), gravel (2 to 15 mm), pebble (16 to 63 mm), cobble (64 to 256 mm) and boulders (>256 mm). Besides the substrate size, embeddedness is also important and ranges from negligible (less than 5 percent of gravel, pebble, cobble and boulder surfaces are covered by fine sediment, grain size <2 mm) to very high, when more than 75 percent of the surface is covered by fine sediment, which is typical for urban waters.

Other important physical habitat features include the following:

* Cover and refuge, provided by boulders, large wooden debris, aquatic vegetation, water turbulence and depth. Such features can provide shelter for fish from predators and physical conditions like fast currents and sunlight.
* Stream-bank and shore conditions, which provide a transition between aquatic and terrestrial ecosystems. Stream-banks in good condition provide cover and refuge for fish. Both natural and anthropogenic impacts may reduce bank vegetation, erosion resistance, structural stability and eventually fish cover value.
* Barriers obstructing fish life cycles (e.g., migration), flow, sediment transport and thermal regime. The degree of disruption depends on the structure height; in urban areas, low to intermediate barriers (1 to 10 m) can be expected in the form of dams forming reservoirs (Bain and Stevenson, 1999).

Heterogeneous physical habitat structure (Bain and Stevenson, 1999) is essential for supporting biodiversity. Complex and varied physical habitats contribute to good performance of fish and benthic communities, and provide habitats for water vegetation, while their simplification reduces biological community performance (Roesner et al., 2005). Straightening banks and enlarging or flattening river bottoms lead to shallow waters, low concentration of dissolved oxygen, elevated water temperature and a lack of shadow and shelter during low flow season. Forced linear flow of water unifies physical conditions within a river and thus diminishes the ecosystem's hydrological heterogeneity.

2.2.3 Biotic interactions

Biotic interactions include such processes as competition, predation, parasitism, feeding, reproduction and disease (Horner et al., 1994). Whenever the natural balance of such processes is disturbed, the biological integrity of the ecosystem is also affected. Habitat simplification significantly limits food sources for the aquatic biota, possibly causing the strength of biotic interactions and the trophic structure of assemblages to undergo irreversible degradation. Changes in primary and secondary production, the disruption of life cycles of native organisms or the introduction of alien species may further impact biological interactions by changing the relationship between predator and prey and other competitive interactions and increasing the frequency of disease (Spence et al., 1996).

2.2.4 Food (energy) sources

To meet organism growth and reproduction requirements, an adequate supply of energy is required (Spence et al., 1996). Thus, physical, chemical and biological processes in the water body and the riparian ecotone must be adequate to produce food resources corresponding to the natural range of abundance (Zalewski et al., 2001). Such requirements are generally poorly understood and often replaced by surrogate requirements of maintaining a corresponding level of primary production and the appropriate riparian vegetation.

2.2.5 Chemical variables (water quality)

Good water quality is an important requirement for habitat integrity. The list of constituents generally includes water temperature, turbidity (suspended solids), dissolved gases (e.g., oxygen), nutrients, heavy metals, selected inorganic and organic chemicals, and pH (Spence et al., 1996).

Urban effluents generally contribute to the increased temperature of receiving waters, particularly in the case of stormwater discharges in summer months when stormwater temperatures are elevated by up to 10°C compared to those in natural water bodies (Schueler, 1987; Van Buren et al., 2000). Discharges of waste heat into receiving waters contribute to reduced concentrations of dissolved oxygen and the replacement of cold-water species by warm-water species.

Both inorganic and organic solids occur in urban waters in high concentrations, contribute to high turbidity and cause many adverse impacts on habitats. Inorganic solids come from streambed erosion, which undercuts stream banks, causes the loss of riparian vegetation and sweeps away habitats (Booth, 1990; Urbonas and Benik, 1995). Eroded materials may be transported to other stream reaches, ponds, lakes and wetlands, with direct and indirect impacts, including interference with photosynthesis, changes in algal productivity, the physical abrasion of fish gills and other sensitive tissues, the blanketing of gravel spawning substrates, the burial of benthic organisms, and influxes of adsorbed pollutants (Horner et al., 1994).

Nutrients (nitrogen and phosphorus) affect the productivity of aquatic systems, and excessive nutrient concentrations, particularly of soluble phosphorus and nitrate nitrogen (NO_3-N), contribute to eutrophication. Some of the eutrophication effects related to human health are addressed in Chapter 8. Additional concerns were reported for ammonia, which is also toxic to fish (Spence et al., 1996).

Other toxic effects in urban waters may be caused by elevated levels of chlorides, heavy metals and trace organic contaminants (Marsalek et al., 1999). Furthermore, high chloride levels may contribute to the mobility of heavy metals (Novotny et al., 1998) and enhance their toxic effects (Marsalek et al., 1999; Rokosh et al., 1997). Both low and high pH levels can be harmful to fish (Spence et al., 1996).

It should be emphasized that the five factors discussed above are not independent, but rather interdependent to various extents. Particularly influential is the flow regime, which effectively governs energy flow, sediment erosion and deposition, and the habitat structure formation or sustainability. Flow regime also affects biotic interaction (by contributing to the sustenance or extinction of some species), energy sources (by impacting on such sources, e.g., by bank erosion), and chemical variables (providing hydraulic transport and the dilution of influents, while contributing to the partition of chemicals in water bodies). Thus, with respect to urban aquatic habitats, flow regime management is particularly important.

Another important issue is the protection of aquatic macrophyte communities. Aquatic macrophyte communities are comprised of emergent, floating and submerged plants. They provide multiple functions for aquatic fauna, such as providing shelter, egg laying sites, nursery areas and food sources for herbivores. Especially dense stocks of floating and submerged macrophytes can have a strong impact on the oxygen concentrations and pH of the water by the processes of photosynthesis and respiration. Increased turbidity levels during flood flows can be deleterious to aquatic macrophytes. Changes that have an impact on macrophyte communities can subsequently affect fauna species that rely on them for food or habitat.

Understanding the abiotic-biotic interactions is important for the assessment of ecosystem quality. The chemical monitoring of waters requires identification of the constituents of critical importance, assessment of their bioavailability, deciding about frequency of sampling and increases in the cost of analyses. Therefore in many cases, bio-monitoring methods, based on the assessment of the ecosystem's biological community, is used to describe water quality in the receiving waters as well as their ecological status (Lafont et al., 2001). This methodology has been successfully extended to urban streams affected by wet-weather discharges (Grapentine et al., 2004), shallow groundwater aquifers providing baseflow to streams and rivers (Lafont et al., 2006), and stormwater control facilities (Datry et al., 2003).

The health of aquatic ecosystems can be characterized by the quality of the general balance between organisms and their environment (Brinkhurst, 1974), which in turn can be characterized by the features introduced in this section. These interactions must be understood, even if this understanding is incomplete, when defining environmental management objectives for aquatic habitats. All major factors controlling habitat structure and performance of the biological community in urban waters are strongly affected by anthropogenic activities. Urban aquatic habitats are also sometimes referred to as effluent-dominated aquatic ecosystems. This makes the challenge of attaining sustainable management of aquatic habitats in urban areas even greater. Moreover, these changes, by impacting the functioning of ecosystems, also reduce their abilities to provide services to human society, which should be one of the priorities in the urban areas.

2.3 BACKGROUND ON AQUATIC ECOSYSTEM FUNCTIONING

Urban aquatic ecosystems are dynamic living systems that grow, in terms of energy through-flow and stored biomass, when exposed to nutrient inputs, but regress when overloaded. The overload appears when the amount of nutrients delivered to the ecosystem in a given time is higher than the amount that can be processes by the kinetics of bio-assimilation processes.

The kinetics of bio-assimilation processes depends on various factors, such as bottom gradient, bottom sediment permeability, water course flow variability, geomorphic variability, the length of the river with no additional local inputs of nutrients, water temperature and oxygenation, and daylight exposure supporting the process of photosynthesis. The quantification of bio-assimilation capacity of ecosystems has a fundamental meaning in urban aquatic ecosystem's management. It allows an adjustment of urban pollutant inputs to the ecosystems characteristics by the distribution of the inflows over time through temporary detention tanks, constructed wetlands or other buffering systems. In cases of limitation of available room, dense urban development, for example, the bio-assimilation capacity can be also increased by the rehabilitation of the water course itself or the adjustment of physical factors controlling the bio-assimilation capacity. This section presents an approach to the bio-assimilation kinetic process assessment.

2.3.1 Conceptual model of aquatic ecosystem functioning

The study of ecology of urban aquatic habitats is complex because of strong synergies between physical (e.g., urbanization of surrounding landscape, habitat's physical modification) and chemical (pollution) stresses, with a mixture of continuous and intermittent (wet-weather) pollution discharges, and the challenge of planning remedial measures under too many constraints imposed by urbanization (lack of sites for locating remedial measures, low feasibility of limiting urbanization and pollution discharges without huge costs, etc.). Consequently, general ecological concepts cannot be easily applied to urban aquatic habitats, and there is a need for more dedicated approaches (Paul and Meyer, 2001). One such an operational approach (conceptual model) is presented here and is based on the Ecological Ambiance System (EASY) proposed by Lafont (2001).

The EASY approach uses a general balance equation to demonstrate that aquatic ecosystems have their own dynamics and cycling capacity that need to be considered in urban water management. Towards this end, biodiversity (BIO) can be used as a relevant broad indicator of the aquatic ecosystem status, and be recognized as a result of biotic and abiotic interactions. Biodiversity can be described here by the ecological ambience (EA), which depends on mineral and organic inputs (IN) and on the ecosystem capacity to cycle such inputs, which is called ecosystem defences (ED) here. Thus, the corresponding balance equation can be written as

$$BIO = k(EA) = f(IN) - g(ED)$$

where k, f and g are functions of arguments EA, IN and ED, respectively.

Thus, the biodiversity evolution (BIO) can be considered as a dynamic balance between the inputs to the system (IN) and the system cycling capacity (ED). The ED factor represents all nutrient cycling and detoxification processes that result from a synergy of biological (metabolic activities of all living organisms), physical (geomorphic context, dynamics of flow exchanges between surface water and groundwater, etc.) and chemical factors (oxydo-reduction potentials, binding of pollutants by the organic matter, speciation of heavy metals, etc).

Important aspects of ecosystem dynamics include ecosystem resistance and resilience capacity with respect to habitat changes. It is evident that resilience and resistance capacities are fundamental ecological factors when dealing with preservation, rehabilitation or remediation actions. Furthermore, resistance and resilience are specific attributes of each ecosystem. Since a full discussion of this subject is beyond the scope of this book, a less comprehensive proposal is made here: a proposition of an applied operational approach.

The ED factor (ecosystem defences) can be considered as an element approximating resilience and resistance capacity assessment. Resilience capacity may be defined as the time required for recovering a balance between ED and IN factors after the implementation of rehabilitation plans. If a long time (greater than a year) is required, the resilience capacity might be considered as low.

Resistance may be assessed by the ability of an ecosystem to maintain its naturally supported biodiversity domain. A given hydrosystem might be considered less resistant than another one if for the same levels of physical and chemical disturbances, it shifts away from its domain while the other one stays in.

The dynamics of the balance equation are depicted in Figure 2.2, which illustrates the changes in biodiversity vs. increasing inputs of organic and mineral substances. There is a general recognition of the shape of the curve in Figure 2.2, and functions k, f and g are specific for individual ecosystems.

The initial rising segment of the curve in Figure 2.2 indicates a slow increase in biodiversity in response to increasing organic and mineral inputs. This is the resistance domain where actual biodiversity adapts to the average natural inputs by bio-assimilation.

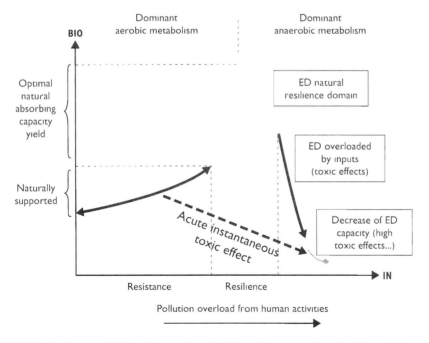

Figure 2.2 The biodiversity (BIO) evolution: dynamic balance between mineral and organic inputs (IN) and defences of the ecosystem (ED)

Source: Lafont, 2001.

Over this domain, with additional chemical inputs from diffuse or point pollution, biodiversity exhibits a rapid increase revealing the development of species that are originally less represented in natural systems until the curve vertex is reached, indicating the highest biodiversity.

The right (falling) part of the curve indicates a decrease of biodiversity resulting from stresses imposed by anthropogenic activities. In the segment corresponding to the upper part of the curve (marked by two-way arrows), the ecosystem can rapidly recover to its original natural state; this is the resilience domain. The region corresponding to the lowest right part of the curve is characterized by extremely adverse effects of toxic substances, such as heavy metals or toxic trace organics. In such conditions, the original ecosystem cannot persist, and the degraded ecosystem cannot rapidly recover. In the case of sudden and highly polluted discharges, such as combined sewer overflows that are disproportionately excessive to the size of the receiving system or accidental toxic pollution spills, biodiversity can suddenly and dramatically diminish, without passing through the intermediate states (acute impacts/changes) indicated by a dashed line. As also shown in Figure 2.2, the associated bio-assimilation processes are dominated by aerobic or anaerobic processes in the domains corresponding to the rising and falling limbs of the curve, respectively.

As mentioned earlier, biodiversity evolution is related to ecosystem defences (ED). Maximum biodiversity corresponds to the availability of adequate ED with respect to chemical inputs. The top of the curve in Figure 2.2 indicates that the yield of nutrient cycling is at its maximum, although inputs are high for the natural aquatic ecosystem in place. To the left of the curve vertex, the yield of nutrient cycling may be low or high, but always greater than the nutrient load. To the right of the curve vertex, pollution loads are increasing, and the effectiveness of the ED factor is decreasing; system defences can be very active, while being overloaded and losing their effectiveness, particularly when toxic substances inhibit nutrient cycling and detoxification processes by living organisms.

In the aquatic ecosystem exposed to intermittent (wet-weather related) organic and mineral inputs, as in urban or suburban environments, biodiversity will move back and forth along the curve. In that case, frequent and significant pollution inputs will result in permanently low biodiversity, as depicted by the right-hand part of the curve. There is no direct correlation between the bio-assimilation capacity and biodiversity in the right-hand part of the curve; the bio-assimilation capacity can be higher in this region, but without corresponding to the original aquatic ecosystem. This situation also results in a decrease in biodiversity leading to an unstable aquatic ecosystem when exposed to the natural flow variability. Such conditions can generate water-related human health risks, which are unacceptable, if not managed.

ED capacities are specific to each system. Low biodiversity and low chemical inputs are typical for harsh environments, like mountain aquatic ecosystems, with small amounts of nutrients (oligotrophic systems) and low mean water temperatures, which slow down nutrient cycling processes. But those ecosystems remain in good health and are used as reference cases. Similarly, naturally rich ecosystems also exist, such as tropical marshes, with complex interactions between food-chains and high water temperatures that stimulate metabolic processes. The highest biodiversity can also be observed in aquatic eco-tones or moderately organically enriched systems (Lafont et al., 2001).

Urban aquatic ecosystems are strongly constrained by anthropogenic stresses but can partly persist using their bio-assimilation capacities (Borchardt and Statzner,

1990). Consequently, the ED factor cannot be ignored and has to be included in management plans for preserving or rehabilitating biodiversity. Towards this end, one can use the conceptual approach presented in this section to develop concrete operational guidance and action measures.

2.3.2 Definition of objectives for preservation or rehabilitation of aquatic ecosystems

The first goal is to define objectives for preservation or rehabilitation plans. Without those objectives, it is not easy to define sustainable management plans. Indeed, there is a risk that remediation or preservation measures might be implemented for individual systems in a stepwise manner, without reference to the global understanding of aquatic habitat management. The main benefit of using a conceptual integrative approach is preventing the dissipation of ideas, means and tools. Figure 2.3 illustrates various proposals of ecological objective criteria.

Pristine or 'very good' ecological status can be easily defined, but its attainment in urban areas is neither probable nor possible. Conversely, an optimal balance between inputs and ecosystem defences might be proposed as an ecological objective that could be selected for preservation or rehabilitation purposes. However, a more realistic objective

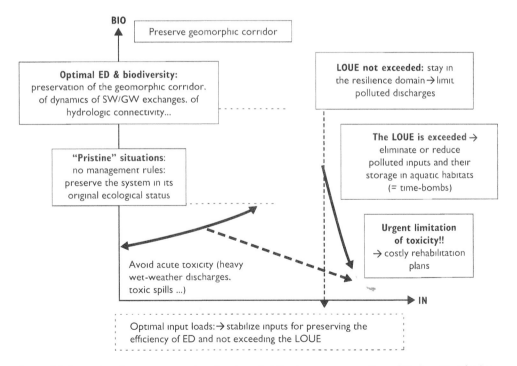

Figure 2.3 Examples of management plans for the rehabilitation or preservation of biodiversity of urban aquatic ecosystems: Optimal ED and biodiversity: optimal adequacy between ED (ecosystem defences) and IN (mineral and organic inputs); LOUE (lowest observed urbanization effect)

Source: Vivier, 2006.

might be to define the 'resilience domain' (see Figure 2.3), in which ED is most efficient. Within this domain, the system is already impacted but quickly rehabilitates when the pollution input is significantly reduced. The term 'quickly' implies that rehabilitation might be achieved over a reasonable period of time, such as one year. Such a time period may be needed to provide evidence that pollution prevention measures have been effective (Vivier, 2006). Of course, this time limit does not include the time needed for building treatment devices or facilities. If after a period longer than one year no improvement is observable, it is necessary to reconsider management plans. For this purpose, an indicator of the Lowest Observed Urbanization Effect (LOUE) must be defined. This corresponds to the lowest detrimental effect in urban aquatic ecosystems, which might be close to the optimal adequacy point between ED and IN, but is located on the right-hand side of the curve (Figure 2.3). The LOUE represents an already impacted situation, which remains within the resilience domain. Starting from a degraded situation and having operated remediation measures for one year, the management plans might be considered as effective if the LOUE is reached or detrimental effects do not persist.

LOUE can be a performance indicator of realistic objectives in aquatic ecosystem's rehabilitation. When the LOUE is significantly exceeded, remediation must be applied in the management strategy (Figure 2.3).

There is then a need to possess suitable tools for assessing the optimal adequacy point, the resilience domain and altered situations. Such tools will differ depending on whether one is considering streams, natural or constructed ponds and wetlands as well as on the issues addressed. Lists of possible physical and chemical changes/impacts as well as of possible bio-indicators are presented elsewhere in two other publications prepared in this series, particularly *Data Requirements for Integrated Union Water Management* and *Urban Water Cycle Processes and Interactions*.

We propose here to assess the optimal adequacy point between the IN and ED capacities, the LOUE, and the resilience domain by integrative biomonitoring tools, such as the sediment quality triad approach (Chapman et al., 1992), where fine sediments dominate, or harmonization systems (Lafont et al., 2001). Optimal adequacy might be obtained when all the selected biological indices, including, for example, those for invertebrate, diatom, oligochaete, macrophyte and fish communities give the same indication of 'good quality,' also referred to as 'good ecological status' (AFNOR, 2000; 2002; 2003; 2004), or when fine sediment quality indices, chemical analyses and toxicity tests detect no alterations (Chapman et al., 1992). If one index detects alterations but others do not; the LOUE can be defined. It is, however, important to consider which index indicates environmental problems and act accordingly. If only the fine sediment index (e.g., IOBS, AFNOR, 2002) detects alterations, then it means the pollution is stored in fine sediment habitats. Remediation measures have to limit pollution storage, and the efficiency of those measures might be assessed by the above mentioned IOBS index. If pollutants are stored in the hyporheic layer without great adverse effects on other habitats, only the biological indicators of this layer, like functional traits, (Lafont et al., 2006; Vivier, 2006) might be used for routine annual checks, etc. Integrative tools have to be applied periodically to check the evolution of the ecosystem. Between those checks, simpler biomonitoring tools targeting a particular habitat might be selected for regular and annual routine checks.

In summary, recognizing that it is impossible to attain pristine aquatic conditions (high or very good ecological status) in urban aquatic habitats, the question arises whether a good ecological status might be achieved and sustained (Weyand and

Schitthelm, 2006) and, in particular, what remedial measures must be planned. A detailed listing and discussion of various possible measures is beyond the scope of this book, but one can propose actions that preserve or enhance the resilience capacity and prevent significantly exceeding the LOUE. Some (but not all) possible management measures are highlighted in Figure 2.3.

2.3.3 Selecting preservation or rehabilitation measures

In this section, the seven main measures for enhancing the aquatic ecosystem resilience capacity are defined as follows:

(1) preserve or rehabilitate the geomorphic corridor of aquatic ecosystems and diversity of aquatic habitats (realistic in suburban areas)
(2) preserve or restore the hydrologic connectivity and the dynamics of hydrologic exchanges between surface water and groundwater
(3) prevent pollutant storage in the ecosystem, either in the hyporheic sub-system or in fine sediments, storage which constitutes a time-bomb when environmental conditions become favourable for pollutant release
(4) avoid excessive pollution discharges with respect to the size of the receiving aquatic habitat
(5) where preserved areas exist (mainly in suburban areas), provide hydrological connections between those areas
(6) permit pollution inputs only if the resilience domain is preserved and the LOUE is not significantly exceeded
(7) check the efficiency of remediation measures every year by specific indices, and every two to three years by integrative methods.

The importance of stream corridors (see Chapters 4, 5 and 9) for stream water quality and biological integrity is well known. The dynamics of hydrologic exchanges between surface water and groundwater greatly stimulate the nutrient cycling (Jones and Mulholland, 2000; Boulton and Hancock, 2006; Breil et al., 2007), but only if the stream-bed is permeable. When the stream-bed is impervious (artificial concrete bed), the transport of pollutants prevails, but pollutants may be stored in downstream areas as soon as conditions favourable for pollutant settling occur. In particular, pollution discharges excessive to the size of the receiving aquatic habitat induce downwelling of surface polluted water and storage of pollutants in the hyporheic layer of gravel streams (Lafont et al., 2006). The reduction of flow velocities in streams or rivers, for example by a dam reservoir or a pond, causes the deposition of polluted sediments originating from upstream reaches. Management practices have to account for this pollutant storage, which is a time-bomb triggered off by the occurrence of environmental conditions favourable for pollutant release (a decrease of the redox potential in fine sediments, upwelling of polluted hyporheic waters to the surface, etc.). Furthermore, remedial actions have to be subject to social and economic considerations. In general, remedial actions are easier to implement in suburban rather than urban aquatic habitats, and in new cities rather than in the older ones.

One final point requires clarification. There is a risk of misinterpretation of the resilience/ED capacity factor. It is not the intention of the remedial action plan to transform receiving ecosystems into wastewater treatment plants, which would not be an acceptable ecological objective for best management practices. The priority must always be given to the pollution source controls or treatment of polluted influents before their entry into the aquatic ecosystem. After the pollution has been reduced or eliminated, then the resilience capacity becomes a stimulating factor for the sustainability of the rehabilitation and preservation of a good ecological quality. However, this approach does not preclude the use of some aquatic habitats for the biological treatment of polluted waters, provided that those habitats might be considered as man-made water bodies and not as reservoirs of biodiversity. These types of habitats do have their place in remediation plans as useful pollution-reducing facilities. Thus, one should promote the idea of managing aquatic habitats devoted to wastewater treatment separately from those devoted to biodiversity conservation.

REFERENCES

AFNOR. 1992. *Essai des eaux: détermination de l'indice biologique global normalisé (IBGN)*. NF T 90–350.

AFNOR. 2000. *Détermination de l'Indice biologique Diatomées (IBD)*. NF T 90–354.

AFNOR. 2002. *Qualité de l'eau – Détermination de l'indice oligochètes de bioindication des sédiments (IOBS)*. NF T 90–390.

AFNOR. 2003. *Qualité de l'eau – Détermination de l'indice biologique macrophytique en rivière (IBMR)*. NF T90–395.

AFNOR. 2004. *Qualité de l'eau – Détermination de l'indice poissons en rivière (IPR)*. NF T 90–344.

Booth, D.B. 1990. Stream-channel incision following drainage basin urbanization. *Water Resources Bulletin*, 26, pp. 407–17.

Booth, D.B. and Henshaw, P.C. 2001. Rates of channel erosion in small urban streams. M.S. Wignosta and S.J. Burges (eds), *Land Use and Watersheds. Human Influence on Hydrology and Geomorphology in Urban and Forest Areas*. Washington DC, AGU, pp. 17–38.

Borchardt, D. and Statzner, B. 1990. Ecological impact of urban stormwater runoff studied in experimental flumes: population loss by drift and availability of refugial space. *Aquatic Sciences*, 52: pp. 299–314.

Boulton A.J. and Hancock P.J. 2006. Rivers as groundwater-dependent ecosystems: a review of degrees of dependency, riverine processes and management implications. *Australian Journal of Botany* 54, pp. 133–44.

Breil, P., Grimm, N.B. and Vervier, A. 2007. Surface water-groundwater exchange processes and fluvial ecosystem function: An analysis of temporal and spatial scale dependency. *Hydroecology and Ecohydrology: past, present and future*. P.J. Wood, D.M. Hannah and J.P. Sadler (eds) Wiley.

Brinkhurst, R.O. 1974. Prospects for forensic biology. Proc. 7th International Conference on Water Pollution Research, Paris, September 9–13, Pergamon Press Ltd., 15 C (1).

Chapman, P.M., Power, E.A. and Burton, G.A., Jr. 1992. *Integrative assessments in aquatic ecosystems. Sediment Toxicity Assessment*, G.A. Burton Jr. (ed.), Lewis Publishers, Chelsea, MI.

Datry, T., Hervant, F., Malard, F., Vitry, L. and Gibert, J. 2003. Dynamics and adaptive responses of invertebrates to suboxia in contaminated sediments of a stormwater infiltration basin. *Archiv für Hydrobiologie*, Vol. 156, pp. 339–59.

Directive 2000/60/EC of the European Parliament and of the Council of 23 October 2000 establishing a framework for Community action in the field of water policy.

Finkenbine, J.K., Atwater, J.W. and Mavinic, D.S. 2000. Stream health after urbanization. *Journal of the American Water Resources Association*, 36(5), pp. 1149–60.

Grapentine, L., Rochfort, Q. and Marsalek, J. 2004. Benthic responses to wet-weather discharges in urban streams in southern Ontario. *Wat. Qual. Res. J.* Canada, Vol. 39, No. 4, pp. 374–91.

Henshaw, P.C. and Booth, D.B. 2000. Natural restabilization of stream channels in urban watersheds. *Journal of the American Water Resources Association*, 36(6), pp. 1219–36.

Horner, R.R., Skupien, J.J., Livingston, E.H. and Shaver, H.E. 1994. *Fundamentals of urban runoff management: technical and institutional issues.* Terrene Institute, Washington DC.

Jones, H.J.B. Jr. and Mulholland, P.J. 2000. *Streams and Ground Waters.* Academic Press, San Diego, CA.

Karr J.R., Fausch K.D., Angermeier P.L., Yant P.R and Schlosser I.J. 1986. Assessing biological integrity in running waters: a method and its rationale. *Illinois Natural History Survey*, Special Publication 5.

Lafont, M. 2001. A conceptual approach to the biomonitoring of freshwater: the Ecological Ambience System. *Journal of Limnology* 60 (Supplementum 1): pp. 17–24.

Lafont M., Camus, J.C., Fournier, A. and Sourp, E. 2001. A practical concept for the ecological assessment of aquatic ecosystems: application on the river Dore in France. *Aquatic Ecology* 35: pp. 195–205.

Lafont, M., Vivier, A., Nogueira, S., Namour, P. and Breil, P. 2006. Surface and hyporheic Oligochaete assemblages in a French suburban stream. *Hydrobiologia* 564: pp. 183–93.

Lapińska, M. 2004. Streams and rivers: Defining their Quality and Absorbing Capacity. Management of streams and rivers: How to Enhance Absorbing Capacity against Human Impacts. M. Zalewski and I. Wagner-Lotkowska (eds) *Integrated Watershed Management – Ecohydrology and Phytotechnology – Manual.* UNESCO IHP, UNEP-IETC, 75–97, pp. 169–88.

Marsalek, J., Rochfort, Q., Mayer T., Servos, M., Dutka, B. and Brownlee, B., 1999. Toxicity testing for controlling urban wet-weather pollution: advantages and limitations. *Urban Water*, Vol. 1, pp. 91–103.

McKay P. and Marshall M. 1993. *Backyard to Bay: the tagged litter report.* Melbourne Water and Melbourne Parks and Waterways.

MacRae, C.R. 1997. Experience from morphological research on Canadian stream: Is control of the two-year frequency runoff event the best basis for stream channel protection? L.A. Roesner, L.A. (ed.) *Effects of Watershed Development and Management on Aquatic Ecosystems, Proceedings of an Engineering Conference*, ASCE, pp. 144–62.

Novotny, V., Muehring, D., Zitomer, D.H., Smith, D.W. and Facey, R. 1998. Cyanide and metal pollution by urban snowmelt: impact of de-icing compounds. *Wat. Sci. Tech.*, Vol. 38, No. 10, pp. 223–30.

Poff, N.L., Allan, J.D., Bain, M.B., Karr, J.R., Prestegaard, K.L., Richard, B.D., Sparks, R.E. and Stomberg, J.C. 1997. The natural flow regime: A paradigm for river conservation and restoration. *BioScience.* 47(11), pp. 769–84.

Paul, M.J. and Meyer, J.L. 2001. Streams in the urban landscape. *Annual Revue of Ecology and Systematics* 32: pp. 333–65.

Roesner, L.A., Bledsoe, B.P. and Rohrer, C.A. 2005. Physical effects of wet weather discharges on aquatic habitats – present knowledge and research needs. E. Erikson, H. Genc-Fuhrman, J. Vollertsen, A. Ledin, T. Hvitved-Jacobsen and P.S. Mikkelsen (eds), Proc. 10th Int. Conf. on Urban Drainage (CD-ROM), Copenhagen, Denmark, Aug. 21–26.

Rokosh, D.A., Chong-Kit, R., Lee, J., Mueller, M., Pender, J., Poirier, D. and Westlake, G.F. 1997. Toxicity of freeway storm water. Goudey, J.S., Swanson, S.M., Treissman, M.D. and Niimi, A.J., (eds) Proc. 23rd Annual Aquatic Toxicity Workshop, Oct. 7–9, 1996, Calgary, Alberta, pp. 151–59.

Schueler, T.R. 1987. *Controlling Urban Runoff: A Practical Manual for Planning and Designing Urban BMPs.* Washington Metropolitan Water Resources Planning Board, Washington DC.

Spence, B.C., Lomnicky, G.A., Hughes, R.M. and Novitzki, R.P. 1996. *An ecosystem approach to salmonid conservation.* TR-4501-96-6057. ManTech Environmental Res.

Services Corp., Corvallis, OR (available from the National Marine Fisheries Service, Portland, OR).

Urbonas, B. and Benik, B. 1995. Stream stability under a changing environment. E.E. Herricks (ed.) *Stormwater Runoff and Receiving Systems: Impact, Monitoring, and Assessment.* Lewis Publishers, Boca Raton, pp. 77–101.

Van Buren, M.A., Watt, W.E., Marsalek, J. and Anderson, B.C. 2000. Thermal balance of an on-stream stormwater management pond. *J. Environmental Engineering,* 126(6), pp. 509–17.

Vivier, A. 2006. Effets écologiques de rejets urbains de temps de pluie sur deux cours d'eau péri-urbains de l'ouest lyonnais et un ruisseau phréatique en plaine d'Alsace. Thesis, L.P. University, Strasbourg (France).

Wadeson, R.A. and Rowntree, K.M. 1998. Application of the hydraulic biotope concept to the classification of instream habitats. *Aquatic Ecosystem Health and Management,* 1, pp. 143–57.

Weyand, M. and Schitthelm, D. 2006. Good ecological status in a heavily urbanised river: is it feasible? *Water Science and Technology* 53 (10): pp. 247–53.

Yoder, C. 1989. The development and use of biocriteria for Ohio surface waters. G.H. Flock, (Ed.), Proc. National Conf. Water Quality Standards for the 21st Century, U.S. EPA, Office of Water, Washington, D.C., pp. 139–46.

Zalewski, M., Bis, B., Frankiewicz, P., Lapinska, M., and Puchalski, W. 2001. Riparian ecotone as a key factor for stream restoration. *Ecohydrology and Hydrobiology,* Vol. 1, No. 1–2. pp. 245–51.

Chapter 3

Strategies, policies and regulations integrating protection and rehabilitation of aquatic habitats in urban water management

Jiri MARSALEK[1], Elizabeth LARSON[2], and Elizabeth DAY[3]

[1] National Water Research Institute, Environment Canada, Burlington ON, Canada L7R4A6
[2] School of Life Sciences Arizona State University, Tempe, AZ, USA
[3] Freshwater Research Unit, University of Cape Town, Cape Town, South Africa

Principles for the protection and rehabilitation of aquatic habitats, and particularly those located in, or affected by, urban areas, have been discussed in Chapters 1 and 2. The main purpose of this chapter is to examine general strategies serving to integrate aquatic habitat issues, and in particular, their protection and rehabilitation, into the general framework of urban water management. The necessity of addressing urban environmental problems in an integrated manner has led to the development of new strategic management concepts, which have been established at various levels and scales, and in different scopes, but which in principle are mutually supportive. With minor exceptions, these concepts are not precisely defined (i.e., multiple, continually evolving, non-contradictory definitions exist in the literature), and that complicates their discussion, which is arranged in the following order: sustainable development (chosen as a guiding principle), followed by the ecosystem approach (as a means of implementation), and performance criteria and indicators for measuring sustainability progress. Finally, our discussion focuses on three management approaches to urban environment components: low impact development, total management of the urban water cycle and soft path for water. All these concepts are first introduced and discussed with reference to aquatic habitats, and finally their applicability to managing urban aquatic habitats is assessed. In the second part of the chapter, policies enhancing sustainability with respect to urban aquatic habitats are briefly presented and discussed. From this perspective, the chapter sets the stage for a discussion of detailed technical aspects of aquatic habitat protection and rehabilitation (Chapters 4, 5) as well as their integration with the city planning and human well-being (Chapters 6, 7 and 8), addressed later in the book.

3.1 STRATEGIES FOR PROTECTION AND REHABILITATION OF URBAN AQUATIC HABITATS

The main strategy adopted here for the protection and rehabilitation of urban aquatic habitats is the principle of sustainable development, which is further discussed with respect to its implementation and criteria for measuring progress towards attaining sustainability.

3.1.1 Sustainable development

The concept of sustainable development was introduced by the Brundtland Commission in 1987 (Brundtland, 1987) and readily embraced by many governments and international organizations. During the following 20 years, more than 250 definitions of sustainable development (and related terms like environmental sustainability) have been proposed, as variations of the original Commission's definition: 'Development that meets the needs of the present without compromising the ability of future generations to meet their own needs.' This definition contextually contains two key concepts: (a) the concept of needs, and in particular the essential needs of the world's poor, to which overriding priority should be given; and (b) the idea of limitations imposed by the state of technology and social organization on the environment's ability to meet present and the future needs.

Among various professions addressing urban development, particularly influential are civil engineers, and it is of interest to note that the American Society of Civil Engineers Code of Ethics (ASCE, 2006) defines sustainable development as a goal, rather than an ideology: 'Sustainable Development is the challenge of meeting human needs for natural resources, industrial products, energy, food, transportation, shelter, and effective waste management while conserving and protecting environmental quality and the natural resource base essential for future development.'

Adoption of the goal of sustainable development then requires implementation approaches and the means of measuring progress towards attaining that goal in the form of various criteria or indicators. As currently practised, the implementation of sustainable development is based on several implementation approaches, among which the ecosystem approach appears to be the most prominent, and contains four pillars: society, environment and economy, supplemented by technical realization. Another implementation approach, less developed with respect to urban environmental practice, is the Natural Step Framework (www.naturalstep.org).

Finally, for individual pillars or sustainability components, the progress towards their attainment is measured by means of various criteria or indicators presented later in this section. A schematic for sustainable development, its implementation and performance measurement is shown in Figure 3.1.

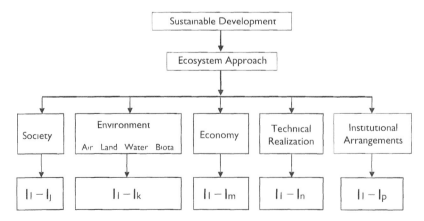

Figure 3.1 Sustainable development: implementation, pillars and progress indicators

3.1.2 Ecosystem approach

The ecosystem approach with reference to environmental planning and management has been articulated in the 1978 Great Lakes Water Quality Agreement (between the United States and Canada; International Joint Commission, 1978) and introduced into environmental planning and management in the Great Lakes Basin in the 1980s (Christie et al.,1986). During the following ten years, many references were published on the ecosystem approach, which can be (and has been) considered as an implementation approach to sustainable development.

There are many definitions of the ecosystem approach in the literature, often with a reference to a particular task or species. As originally introduced in environmental planning and management, the ecosystem approach emphasized the need to attribute equal importance to three basic entities (also referred to as pillars): society, (living) environment and economy. This system of three pillars has been expanded by some authors to five pillars by considering implementation issues and including institutional arrangements (UNDSD, 2005) or technical realization (Balkema et al., 2002; Ellis et al., 2004). One could argue that the institutional arrangements are included in the societal pillar, but in all implementation efforts, the need for strengthening institutional arrangements at all levels (the Johannesburg Plan of Implementation, for example UNDSD, 2005), or developing social and institutional capacity (see Section 3.1.3; Yale and Columbia University, 2005) should be acknowledged. Thus, an all-encompassing approach would include five pillars, as shown in Figure 3.1. Finally, a working definition of the ecosystem approach has been adopted here as a comprehensive regional approach that integrates ecological protection and rehabilitation with human needs to strengthen the fundamental connection between economic and social prosperity and environmental well-being (Ramsar, 1998).

The essential components and key concepts of the ecosystem approach are summarized by an Environment Canada Task Group (1995) as follows:

- The ecosystem approach recognizes that a system's resistance and resilience to change set limits to human activities. This in turn requires the identification of stressors to the system, likely responses and indicators of those responses with full consideration of time lags and cumulative effects. The definition of the health and integrity of the system is based at least in part on these stress/response relationships, and the definition of sustainability of the system in terms of these indicators.
- The ecosystem approach highlights the need for a more comprehensive and interdisciplinary methodology for environmental research, planning, reporting and management, which includes incorporating both the social and physical sciences.
- Recognizing that all the components of an ecosystem (physical, chemical, biological and human) are interdependent, and that a stress on one component may affect many others, the ecosystem approach requires that resources be 'managed as dynamic and integrative systems rather than as independent and distinct elements. Its practice means that all stakeholders understand the implications of their actions on the sustainability of ecosystems' (Wrona and Cash, 1996).
- The dynamic and complex nature of ecosystems requires that the ecosystem approach must also be flexible, adaptive and able to cope with complexity.
- Recognizing that all elements of the system should be planned in unison, water, land use, air and natural resource concerns should be balanced with issues

of human activities and economic development in order to move towards an ecologically sustainable socio-physical system. This requires a more holistic orientation towards the management of environmental and socio-economic considerations.

There are examples of applications of the ecosystem approach to environmental planning and management in the literature; particularly from the Great Lakes Basin, where remedial action plans were developed for areas of concern. Among the fourteen impairments of beneficial uses in the areas of concern, half deal with issues directly related to fish habitats, including their loss (IJC, 1988). Further scientific underpinning for ecosystem management can be found in Zalewski (2000), and practical and systematic measures for watershed management were presented by Zalewski et al. (2004).

3.1.3 Measuring progress towards sustainability: Criteria and indicators

Both terms for performance measurement, criteria and indicators, are often used interchangeably in the literature, but Sahely et al. (2005) make a distinction: indicators measure sustainability or the state of environment by considering a manageable number of variables or characteristics (McLaren and Simonovic, 1999), while sustainability criteria are yardsticks (i.e., goals), against which sustainability indicators are measured. Thus in terms of quantified measurements of progress towards sustainability, indicators are more appropriate and have been adopted, for example, by the United Nations Division for Sustainable Development (UN DSD, 2006). In the following section, several examples of sustainability indicators are presented.

3.1.3.1 UN Division for Sustainable Development indicators (2006)

These indicators are rather broad and cover a spectrum of fourteen categories of indicators: poverty; governance; equity; health; education; demographics; atmosphere; land; oceans, seas and coasts; freshwater; biodiversity; economic development; global partnership; and consumption and production patterns. This list is currently under review, but reflects recent improvements with respect to: (a) emphasis on measuring development progress, (b) a switch from the indicators along the four 'pillars' (social, economic, environmental and institutional) to multidimensional indicators, and (c) the additional classification of indicators into core and non-core categories. In total, forty core indicators were identified as well as thirty-nine non-core indicators, and thirteen were listed in the to-be-developed category. Examples of such indicators are presented in Table 3.1.

3.1.3.2 Environmental Sustainability Index (ESI) Yale University and Columbia University 2005

The ESI was proposed by the Yale Center for Environmental Law and Policy at Yale University and the Center for International Earth Science Information Network at Columbia University. The ESI integrates seventy-six data sets by tracking natural endowments, past and present (or expected) pollution levels, environmental management efforts, and the capacity of the society to improve its environmental performance into twenty-one indicators of environmental sustainability. Such indicators then allow for the comparison of environmental issues falling into five categories: the integrity of

Table 3.1 Examples of UNDSD indicators of sustainable development with potential direct effects on aquatic habitat

Theme	Sub-Theme	Indicator	Type of Indicator
Poverty	Sanitation	Proportion of population with access to improved sanitation	C
		Wastewater treatment level	NC
	Drinking water	Population with access to safe drinking water	C
Land	Land use and status	Land use indicator	TBD
		Land degradation	TBD
	Agriculture	Arable and permanent cropland	C
		Efficiency of fertilizer use	NC
		Use of agricultural pesticides	NC
	Forests	Forest area as a percent of land	C
	Desertification	Land affected by desertification	C
Oceans, seas and coasts	Coastal zone	Algae concentration in coastal waters	C
		Percentage of total population living in coastal areas	NC
		Coastal pollution	TBD
	Fisheries	Proportion of fish stock within safe biological limits	C
Freshwater	Water quantity	Annual withdrawal of ground and surface water as percent of renewable water	C
	Water quality	BOD in water bodies	NC
Biodiversity	Ecosystem	Coverage of protected areas by biome and species	C
		Management effectiveness of protected areas	TBD
		Fragmentation of habitat	TBD
	Species	Trends in the abundance of selected key species	C
		Assessment of threatened species	NC
Consumption and production patterns	Material consumption	Intensity of material use, total and by sector	C
	Energy use	Annual energy consumption, per capita, total and by sector	C
	Waste generation and management	Generation of hazardous waste	C
		Waste treatment and disposal by method of treatment (recycled, incinerated, landfilled)	C
	Transportation	Fuel use by distance of passenger transportation	C

C - Core, NC - Non-Core, TBD - To be developed. Source: from UNDSD, 2006.

environmental systems, reducing environmental stresses, reducing human vulnerability to environmental stresses, societal and institutional capacity to respond to environmental challenges, and global environmental stewardship. Even though this index was specifically developed for national environmental stewardship, many of its elements are applicable to smaller-scale environmental systems. For example, Marsalek (2005) suggested using the ESI for assessing the sustainability of municipal stormwater management programmes that have a direct bearing on urban aquatic habitats by applying the following five indica-tors specific to stormwater management and corresponding to the ESI five categories:

(1) Integrity of the environmental systems and their beneficial uses (aquatic ecosys-tems and terrestrial resources) – stormwater management should contribute to the adequate protection of these systems with respect to their biodiversity (e.g., by pro-tecting or creating habitat), land resources (limiting the extent of degraded lands), protecting water quality in the receiving waters for maintaining or upgrading benefi-cial uses (e.g., fishing and swimming), and preserving water balances of sub-water-sheds (including discharges of runoff, affecting the fluxes of pollutants and sediment).
(2) Reducing environmental stresses by the means of: (a) public education and partici-pation; (b) source controls for pollutants exposed to stormwater in urban areas, including controls for fertilizers and pesticides, the wise use of road salt, substitu-tion of building materials, and reduced traffic; (c) enforcement of sewer ordin-ances (to prevent illicit discharges); (d) good housekeeping practices; (e) control of pollution by construction activities; and, (f) adequate and timely maintenance.
(3) Reducing human vulnerability – by preventing flooding or runoff ponding in urban areas, reducing flood damages, and reducing exposure to chemicals, pathogens or disease vectors.
(4) Social and institutional capacity – this category includes three indicators, dealing with environmental governance (i.e., the availability of institutions or agencies to plan, design and operate stormwater management programmes; existence of legis-lative support), science and technology (data collection/assessment/research pro-grammes; and the availability of innovative products and services), and private sector responsiveness (e.g., with respect to corporate stewardship, development of control measures and technologies).
(5) Regional/global stewardship – includes participation in national/international environmental agreements and programmes, limiting pollutant exports and runoff from the jurisdiction, and controlling greenhouse gas emissions (e.g., by appropri-ate choices of the types of facilities and constructions materials).

3.1.3.3 Pillar-based sustainability criteria

Several other sets of sustainability criteria have been reported, for example, by Balkema et al. (2002) and Ellis et al. (2004), and refer to relative performance rating for the previously mentioned pillars of the ecosystem approach: the society, (living) environment, economy, and technical implementation. The specific criteria used are selected by the stakeholders, and serve for comparing the environmental sustainability of various project alternatives. Where useful, weights may be ascribed to individual criteria. The option with the highest rating is adopted for implementation. A list of commonly used pillar-based criteria can be found in Table 3.2.

3.1.4 Urban environmental management approaches

Urban impacts on receiving waters and their ecosystems have been recognized in the past thirty to forty years, and new approaches to reducing or mitigating such impacts

Table 3.2 Sustainability criteria

Social performance	Economic performance	Environmental performance	Technical performance
Public health	Affordability	Surface water quality & quantity	Durability
Values and beliefs	Ability to finance	Groundwater quality & quantity	Ease of construction
Cultural resources	Ability to maintain	Ecosystem protection	Withstands shock loads
Public involvement: awareness and participation	Ability to sustain	Air, soil quality	Flexibility
Aesthetics Community development	Economic development	Energy use Biodiversity	Maintenance Reliability, security
		Pollution prevention	On-site solution

Source: Balkema et al., 2002; Ellis et al., 2004.

Table 3.3 Characteristics of Urban Environment Management Approaches

Management Approach	Environmental Component			
	Air	Land	Water	Biota
Low Impact Development	O	+ + +	+ +	+
Total management of the Urban Water Cycle	O	+ +	+ + +	+
Soft path for water	O	+ +	+ +	+

Legend: O – minimal priority, minimal (or not measurable) effect/benefit; + – Low priority, effect, benefit; + + – Intermediate priority, effect, benefit; + + + – High priority, effect, benefit.

have been proposed. Only recently these approaches gained a broader acceptance under such terms as low impact development (LID)[1], total management of the urban water cycle, or soft path for water. Their basic characteristics are listed in Table 3.3.

3.1.4.1 Low impact development (LID)

LID was introduced during the 1980s in the Northeastern USA, but the essential features of LID had been already applied ten years earlier in Woodlands, Texas (starting in 1974). LID is generally defined as a comprehensive, landscape-based approach to sustainable urban development encompassing strategies to maintain existing natural systems and their hydrology and ecology. In practice, typical LID literature focuses on preserving site hydrology and water balance, which is quite challenging in urban developments with many impervious elements. Consequently, preserving a site's natural hydrology and ecology is more of a goal than an attainable objective, because research

[1] This term is common in the American literature; similar terms include water sensitive urban design, used in Australia, or sustainable urban drainage systems (SUDS), used in the UK.

indicates that when as little as 10% of the watershed is developed, the receiving stream becomes clearly impacted by this development (Horner et al., 1994). Thus, some authors question how low is 'low' (Strecker, 2001) and suggest that a 'lower' impact development might be a more accurate term.

The basic LID principles can be summarized as follows:

- Minimizing development impacts by conserving/restoring natural resources and ecosystems and maintaining natural drainage in site planning; minimizing clearing and grading; and reducing imperviousness (particularly the so-called directly connected impervious areas, e.g., by building narrower roads and using pervious parking lots)
- Maintaining, recreating or enhancing distributed detention and retention storage on sites by using open drainage swales, flat slopes, rain gardens (bio-retention) and rain barrels
- Maintaining the pre-development time of concentration and travel times by strategically routing runoff flows
- Encouraging property owners and drainage system operators to use effective pollution prevention measures and maintain all management measures
- Managing and treating stormwater close to the sources (particularly for frequent events) and avoiding the excessive export of runoff flows and associated pollutants or materials
- Emphasizing simple, non-structural low-cost management measures (policies, drainage from impervious areas onto pervious areas, stormwater infiltration and reuse, public education and involvement)
- Creating multifunctional landscapes, combining stormwater management areas with recreational areas and wildlife habitat
- Maintaining all landscape features and stormwater management facilities
- Achieving the above actions without significantly increasing urban sprawl (Boston MAPC, 2006).

The need to keep urban sprawl under control cannot be overemphasized, because sprawl reduces environmental sustainability by using much more land per person, thereby consuming new land for urban development, increasing road construction and maintenance as well as automobile travel, thereby contributing to air pollution by increased consumption of gasoline and other non-renewable products; increased water, sewer, and other utility line construction and their long-term maintenance; lowering municipal tax revenues in relation to the cost of providing infrastructure and related services; inducing the relocation of people from central parts of cities and inner suburbs; increasing the export of sediment, nutrients and other chemicals from former rural areas; increased lengths of streams/rivers and other natural water bodies in direct proximity to urban populations, with adverse consequences for their habitats; and, decreasing wildlife habitats.

3.1.4.2 Total management of the urban water cycle (TMUWC)

TMUWC is another concept addressing integrated water management in urban areas (Lawrence et al., 1999). The concept demonstrates the connectivity of urban water components, their interdependence on human activities and the need for integrated management. The main features of TMUWC can be summarized as follows:

- Re-use of treated wastewater as a basis for disposing of potential pollutants, or as a substitute for other sources of water supply for non-potable uses
- Integrated stormwater, groundwater, water supply and wastewater management as the basis for economic and reliable water supply; environmental flow management (deferment of infrastructure expansion, return of water to streams); urban water-scape/landscape provision; substitute sub-potable sources of water (wastewater and stormwater reuse); and, protection of downstream waters from pollution
- Water conservation-based (demand management) approaches, including more efficient use of water (water saving devices, irrigation practices); substitute landscape forms (reduced water demand); and substitute industrial processes (reduced demand and water recycling).

Many of these measures have been practised individually in the past, but without the full understanding of linkages among the various components and the implications of these practices for long-term quality of groundwater, soils and environmental flows.

3.1.4.3 Soft path for water

The soft path for water concept has appeared in recent literature on water policies. It is not explicitly defined in the literature, but rather is described by some of its features and differences from the conventional 'hard' path in water management relying on centralized infrastructures. A working definition could be paraphrased from the information offered by Wolf and Gleick (2003) and the website of the Rocky Mountain Institute (www.rmi.org): the soft path for water management is a management approach based on greatly increased efficiency in end water use, incorporating precise management systems to avoid system losses and matching system components to the exact quantities and qualities required for appropriate classes and locations of end use. The soft path may employ both centralized and decentralized infrastructure, efficient technologies and human capital. It focuses on improving the efficiency and productivity of water use, rather than seeking new water supplies. Towards this end, it delivers water services matching users' needs and engages local stakeholders in this process.

Soft path challenges the current patterns of water use and adopts the following four principles:

- Treat water as a service rather than an end in itself
- Make ecological sustainability a fundamental criterion
- Match the quality of water supplied to that needed by the end-user
- Plan from the future back to the present (Brandes and Brooks, 2006).

The principles of the soft path for water management relate quite well to the objectives set for managing aquatic habitats in urban areas or areas affected by urbanization. Specifically, water provides an important service with respect to the habitat by supporting aquatic life. Ecological sustainability, referring to various levels of organisms in water bodies, preserves biodiversity and supports healthy aquatic habitats. Matching the water quality to that needed in support of aquatic life also serves to support healthy aquatic communities. Finally, planning from the future, i.e., attaining good aquatic habitats with healthy biological communities, helps define the levels of stress or the degree of rehabilitation needed with reference to the existing water bodies. Other environmental management approaches to enhancing urban aquatic habitats are discussed in Chapter 6 dealing with ecohydrology (Zalewski, 2000). While this book focuses on urban aquatic habitats, the need to address the entire watershed cannot be overemphasized, and specific measures serving such a goal have been presented elsewhere (Zalewski et al., 2004).

In summary, none of the contemporary progressive approaches to urban water resources management specifically addresses the issues of protecting the urban aquatic habitat, but all of them emphasize actions, which should support the habitat protection through such measures as maintaining the site water balance in urban developments; reducing water use (and its inefficiency) by conservation, water reuse and water saving landscape and vegetation and technologies; integration of the management of individual water service components; and enhancing public education and participation in these efforts.

3.2 REGULATIONS AND POLICIES DRIVING THE IMPLEMENTATION OF STRATEGIES FOR URBAN AQUATIC HABITAT'S PROTECTION AND REHABILITATION

New strategies for managing urban waters and protecting urban aquatic habitats are implemented by introducing specific policies and guidelines in various countries, or blocks of countries (e.g., the European Union). It is generally recognized that water policies are difficult to transfer among countries with different levels of development, socio-economic conditions, and environmental challenges. Consequently, only a general discussion of the types of policies used in various countries to protect and restore aquatic habitats is presented and further documented by examples from a number of countries.

The issues of aquatic habitat are regulated in various countries by legislation concerned with water quality, fish habitat, aquatic habitat, ecological status and the protection of species, biodiversity and natural resources. The presentation of material starts from habitat-specific policies and moves towards more general policies.

3.2.1 Approaches focusing on aquatic habitat protection

Two examples of policies focusing on fish habitat protection are the Fisheries Management Act 1994 (Government of New South Wales, Australia) and the Canada Fisheries Act (CFA) (NSW Fisheries, 1994; DFO, 2005). Both acts aim to promote ecologically sustainable development by protecting fishery resources and fish habitats. To this end, the CFA prohibits the discharge of deleterious substances to fish-inhabited waters (tested by a 96-hour rainbow trout bioassay) and prohibits the destruction of fish habitats. According to the act: 'No person shall carry on any work or undertaking that results in the harmful alteration, disruption or destruction of fish habitat.' While this wording contains fairly sweeping powers, there are challenges in applying the act, particularly when dealing, in a piecemeal fashion, with individual projects potentially causing minor (and often non-measurable) impacts, but the accumulation of such impacts in the whole catchment may have significant implications for fish habitats.

Habitat issues are often addressed in connection with stormwater management controlled by specific policies and regulations that may or may not cover all the major factors affecting stream habitat and biological integrity: flow regime, physical habitat structure, chemical fluxes, energy (food sources), and biotic interactions. To this end, more attention must be paid to stormwater treatment trains (Schueler, 1987), starting with non-structural measures (including low-impact development planning, low catchment imperviousness, preservation of natural drainage, pollution control at the source through product substitution); proper selection, sizing and construction of structural best management practices (BMPs), and maintenance and upgrading of the older ones; preserving on-site water balance and enhancing stream channel stability; and providing in-stream aquatic habitat rehabilitation and restoration (e.g., developing and protecting riparian vegetation, stream buffers, rehabilitating floodplains, etc.). In many instances, this involves a catchment that has been undergoing urbanization for some time. This will involve the rehabilitation of existing in-stream conditions, control of sources of runoff flows and pollution in the catchment and control of catchment discharges.

Where a new habitat is created, e.g., by constructed wetlands or stormwater ponds, the eco-toxicological status of the polluted sediment accumulating in BMPs needs to be assessed (Bishop et al., 2000 a, b). Since BMPs are designed to remove suspended solids from stormwater, large quantities of sediment, often polluted by anthropogenic sources, accumulate in BMP facilities (Marsalek et al., 2006). Besides sediment, chloride originating from winter road maintenance also accumulates in BMPs and reduces biodiversity. Sediment beds represent a poor habitat for benthic organisms; indeed, often only pollution-tolerant species survive in such habitats. The lack of benthic organism populations, or their contamination, then also adversely affects fish communities, for which benthos is the source of energy.

3.2.2 Approaches focusing on water quality protection

Two examples of aquatic habitat protection through regulations focusing on water quality are featured here: The South African National Water Act and the US Environmental Protection Agency (EPA) approach. The South African National Water Act recognizes the aquatic ecosystem as both a resource and a legitimate user of the resource in its own right. Target water quality guidelines have been developed, which apply to the quality of water in rivers, and are used in the derivation of Resource Directed Measures (RDMs). RDMs focus on the water resource itself, and their main purpose is the setting of clear objectives for a desired level of protection for a water resource and satisfaction of the water quality requirements of water users within reasonable limits (DWAF, 2000). These measures require that water resources be grouped into different protection classes, with each class representing a different level of protection. By contrast, source-directed measures focus on the sources of impacts on water resources, from both point and diffuse sources. They are aimed at controlling the generation of waste at source (DWAF, 2000).

The requirements for ensuring the long-term use of water resources mean that an approach based on both resource- and source-directed measures is needed. To give effect to the measures, a person may only discharge waste of a certain quality into a water resource, such as a river. DWAF (2000) outlines a process for determining appropriate waste standards, involving both the above water resource and user-protection aspects (resource-directed measures) as well as technological (what can be achieved technologically and economically, taking into account the associated social and economic impacts) and source-directed aspects.

A somewhat similar approach is practised in the United States, where the main statutes concerning aquatic habitats derive from the Clean Water Act (CWA), created in 1972, which regulates the discharge of pollutants into the nation's waterways (Copeland, 2002). One part of the CWA of ongoing concern is Section 404, which regulates the discharge of dredged or fill materials into American waters, including wetlands. This act should, in theory, protect all wetlands during urbanization, but when developers can demonstrate that no practical alternative exists that is less damaging, then mitigation may be used to create a new wetland elsewhere to compensate for the loss. Another aspect of the CWA that is important to urban environmental managers is the 'Stormwater Phase II Rule', part of the National Pollutant Discharge Elimination System (NPDES), Section 402 of the CWA. This ruling covers stormwater discharges from municipal separate storm sewer systems and from construction activities, and requires the development and implementation of best management practices (BMPs) to eliminate illicit discharges, control construction site runoff and prevent pollution.

It should be noted that the US Environmental Protection Agency (EPA) and state regulations affecting urban aquatic habitats are primarily focused on protecting or improving water quality. Aside from Section 404, there are no federal regulations directly protecting aquatic habitat structures. One explanation for this is the difficulty in establishing specific measures that would accommodate the huge variety of ecosystem types and uses throughout the entire US. Divisions within the EPA, such as the National Estuary Program, and various state programmes have connected water quality issues to maintaining biota and ecosystem functioning and thus have developed conservation and management plans.

The National Environmental Protection Act (NEPA) is a programme that may indirectly deal with urban aquatic habitat protection. Enacted in 1969, NEPA requires all federal agencies to review their impact on the environment. New projects must file an environmental impact statement and present research-supported information on any irreversible effects and costs of, as well as alternatives to, the proposed action. The impact statements are filed before any permitting processes to ensure that the permitting agencies know the environmental consequences of their actions. Many states have implemented their own acts, requiring state agencies to do the same.

At the local level, municipalities have the ability to identify particular habitats of concern and take measures to ensure their protection. Zoning regulations are a common approach at this level. For example, riparian buffer width is regulated in the Portland, Oregon metropolitan area. However, many of the alterations to, and destruction of, urban aquatic habitats happened years or decades before environmental awareness and science reached their current levels of development, so CWA Section 404 and zoning laws only apply to new development. For older areas, there are several government and private/NGO funding options for restoration and management.

Special projects are, by definition, aimed at addressing specific problems or types of ecosystems and thus tend to be locally implemented, regardless of the source of funding. Thus, because problems in urban aquatic ecosystems are often severe and widespread, only a small number of habitats receive significant monetary attention. For example, the

Urban Rivers Restoration Initiative, a joint project between the EPA and the US Army Corps of Engineers (USACE), was only able to fund river cleanup and restoration on eight rivers nationally since 2002. USACE is a frequent collaborator in similar projects, sharing costs with municipal, state, and federal agencies for environmental engineering projects.

US federal funding of environmental projects can be dictated by national politics and decisions made far from the location of the problem, especially because the heads of both the EPA and USACE are appointed by the US President. This is one of the reasons that more local, community-based options may be more enduring and appealing to local stakeholders, although, of course, local politics and economic dynamics may influence outcomes as well. One example of non-federal project funding is California's Urban Streams Restoration Program, which has sponsored numerous programmes, such as cleanups, invasive plant eradication, bioengineering bank stabilization, channel reconfiguration, etc. Frequently, awards have been made for projects co-sponsored by both municipal authorities and citizen action groups. Non-profit organizations, such as the Center for Watershed Protection and the Rocky Mountain Institute, also play an important role in organizing stakeholders, conducting research and project management.

3.2.3 Integrated regulations and policies addressing aquatic habitats

One of the most comprehensive and integrated water policies is the European Union's Water Framework Directive (WFD – Directive 2000/60/EC). The WFD includes recommendations for the assessment and protection of surface and groundwater, but in general, it aims for the assessment and protection of the aquatic ecology, and specifically the protection of unique and valuable habitats, drinking water resources and bathing water.

The WFD establishes links to the ecosystem approach by adopting targets (subject to various qualifications) for ecological protection of all surface water body types, including rivers, lakes, transitional waters and coastal waters. It recommends Member States to take action in order to achieve 'good ecological status' of the natural water bodies by 2015. Ecological status defined in Annex V of the WFD provides a detailed classification of quality elements describing both the biological community (including flora and fauna), as well as the hydrological and chemical characteristics of the ecosystem.

For biological quality, no absolute standards can be set across the community, so the controls are specified to permit a small departure from the biological community expected in minimally anthropogenically impacted conditions, so-called reference conditions. These reference conditions are specified for each water body type and describe ecosystems representing relatively unmodified habitats with high values of the hydromorphological and physicochemical quality elements. The selection of those ecosystems allows for establishing type-specific biological reference conditions. Procedures for identifying 'good ecological status' for a given aquatic habitat, together with chemical or hydro-morphological standards needed to achieve this status, are provided. Good chemical status is defined with respect to compliance with all the quality standards established for chemical substances by the European Union.

Urban aquatic habitats, which are often greatly impacted, may be classified according to the WFD as a separate class of the so-called 'heavily modified water bodies' (i.e., surface water bodies, which as a result of physical alterations by human activities are substantially changed in their character) or even as 'artificial water bodies' (surface water bodies created by human activities). They very often cannot be restored to the highest quality and achieve 'good ecological status'. In that case, the WFD recommends using the concept of Maximum or Good Ecological Potential (MEP/GEP), which is defined as the potential for the development of chemical, physical and biological processes within realistic boundaries. The quality elements applicable to artificial and heavily modified surface water bodies shall be those applicable to whichever of the four natural surface water categories discussed in the WFD most closely resembles the heavily modified or artificial water body concerned.

The WFD also allows the designation of specific protection zones within the river basin that must meet different, more stringent objectives required for 'other uses'. Other parts of the WFD deal with groundwater chemical status (adopting a precautionary approach) and quantitative status, allowing only that portion of the recharge to be abstracted that is not needed under the ecological requirements.

Finally, the WFD uses a combined approach, applying all existing technology–driven, source-based controls in the first step. And where the source controls are not sufficient to meet the overall objective of good status of all waters, it applies additional measures. The WFD emphasizes public participation by involving stakeholders in river basin management planning and enhancing enforceability, while introducing true cost pricing for all water services. So far, the experience with the WFD is limited, with the deadline for achieving good ecological status set by 2015.

The EU has also been actively involved in the protection of natural habitats, mostly through the Habitats Directive (Council Directive 92/43/EEC of 21 May 1992 on the conservation of natural habitats and of wild fauna and flora) and the related Birds Directive (Council Directive 79/409/EEC on the conservation of wild birds). The Habitat Directive creates a European ecological network of 'special areas of conservation' called Natura 2000, protects animal and plant species of special interest for the community and integrates nature protection requirements into other EU policies, such as agriculture, regional development and transport. The Natura 2000 network often comprises landscape closely associated with aquatic habitats like river valleys. Although in many cities aquatic habitats have been heavily degraded, there are also examples of reaching a compromise between urban development and biodiversity protection. An example of the Natura 2000 site with endangered biota, located in a highly urbanized area of the Adige Basin (Italy), is provided in Chapter 9 of this book. Another example is provided by the attempts undertaken in the riparian wetlands still existing after the regulation of the Danube River, at the northeastern outskirts of Vienna (see Chapter 9). This area is not only protected by the Natura 2000 network, but has also a status of a UNESCO Biosphere Reserve, a RAMSAR area and a Landscape conservation area following an act of the Municipality of Vienna, as a part of the Danube National Park.

Protecting biodiversity has been set as a top priority in the Sixth Environmental Action Plan (EAP), Environment 2010: Our Future, Our Choice, which sets out the EU's environmental policy agenda until 2012. The European Community is also one of the signatory parties of the Convention on Biological Diversity (CBD).

3.2.4 Additional observations concerning developing countries

Aquatic habitats and their protection or restoration are important in all countries, regardless of their level of economic development. While in the developed countries drivers behind the associated policies appear to be biodiversity and environmental stewardship, in developing countries, one of the important drivers is protecting fishing as a major source of food and a provider of employment and economic benefits, particularly in rural areas (Sverdrup-Jensen, 1997). Even though inland capture fisheries probably represent no more than 10% of the world's supply of fish for human consumption, in specific areas, it is an important source of food that needs to be protected, particularly in the case of low income groups and communities facing food insecurity. With respect to increasing inland fisheries, the need for integrated management has been expressed, particularly by preventing aquatic habitat degradation (or by undertaking habitat rehabilitation), avoiding flow alterations, and preventing pollution. These issues require urgent attention in the assessment of proposed development activities and should eventually become entrenched in the environmental protection legislation of developing countries, as demonstrated in Guyana by Lakhan et al. (2000).

1.2.5 Emerging challenges

Studies addressing existing aquatic habitats and their relation to the stormwater practices indicate that the mere presence of best management practices (BMPs) does not guarantee good habitat conditions in the receiving waters (Urbonas and Jones, 2002; Roesner et al., 2005). Many early BMPs were either designed for limited objectives (e.g., just peak flow shaving), or under-designed (sometimes in the view of continuing urbanization or climate variations and changes), or not maintained adequately as they near the end of their design life. Thus, when addressing the relationship between the habitat quality or fish community performance and the stormwater management practice, a deeper understanding of BMP systems and their performance on the catchment/receiving-water-body scale is needed. Furthermore, recognizing that catchments and their ecosystems are highly dynamic systems subject to time-varying stressors, it is of interest to consider the changes in stressors and their potential impact on catchments and their ecosystems. In this context, perhaps the most significant challenges include growing urban populations, climate change, and aging or outdated infrastructure. These concerns are particularly applicable to urban flood protection and drainage infrastructure. Some of the early BMPs for stormwater management are now 30 to 40 years old and may require substantial upgrading to meet the design objectives of contemporary stormwater management. Adopting a conservation/sustainable approach to urban development helps reduce, but does not prevent entirely, adverse effects of urbanization and their implications for urban aquatic habitats (Horner et al., 2002).

REFERENCES

American Society of Civil Engineers (ASCE). 2006. *Code of Ethics*. (http://www.asce.org/inside/codeofethics.cfm).

Balkema, A.J., Preisig, H.A., Otterpohl, R. and Lambert, F.J.D. 2002. Indicators for the sustainability assessment of wastewater systems. *Urban Water*, Vol. 4, pp. 153–61.

Bishop, C.A., Struger, J., Barton, D.R., Shirose, L.J., Dunn, L, Lang, A.L. and Shepherd, D. 2000a. Contamination and wildlife communities in stormwater detention ponds in Guelph and the Greater Toronto Area, Ontario, 1997 and 1998. Part I – Wildlife communities. *Water Qual. Res. J. Canada*, Vol. 35, No. 3, pp. 399–435.

Bishop, C.A., Struger, J., Shirose, L.J., Dunn, L. and Campbell, G.D. 2000b. Contamination and wildlife communities in stormwater detention ponds in Guelph and the Greater Toronto Area, Ontario, 1997 and 1998. Part II – Contamination and biological effects of contamination. *Water Qual. Res. J. Canada*, Vol. 35, No. 3, pp. 437–74.

Boston Metropolitan Area Planning Council (MAPC), 2006. *Massachusetts Low Impact Development Toolkit*. http://www.mapc.org/regional_planning/LID/LID_into2.html

Brandes, O.M. and Brooks, D.B. 2006. *The Soft Path for water in a nutshell*. POLIS Project, University of Victoria, Victoria, BC, May.

Brundtland, G. (ed.) 1987. *Our common future*. The World Commission on Environment and Development, Oxford University Press, Oxford, UK.

Copeland, C. 2002. Clean Water Act: A Summary of the Law. Congressional Research Service, The Library of Congress, Order Code RL 30030, Washington DC.

Council Directive 79/409/EEC of 2 April 1979 on the conservation of wild birds.

Council Directive 92/43/EEC of 21 May 1992 on the conservation of natural habitats and of wild fauna and flora.

Christie, W.J., Becker, M., Cowden, J.W. and Vallentyne, J.R., 1986. Managing the Great Lakes Basin as a home. *J. Great Lakes Res.*, Vol. 12, No. 1, pp. 2–17.

DFO (Department of Fisheries and Oceans) 2005. Administration and enforcement of the fish habitat protection provisions of the Fisheries Act. Annual report to the Parliament, DFO, Ottawa, Ontario, Canada (available at www.dfo-mpo.gc.ca).

Directive 2000/60/EC of the European Parliament and of the Council of 23 October 2000 establishing a framework for Community action in the field of water policy

DWAF (Department of Water Affairs and Forestry). 2000. *Resource Directed Measures for Protection of Water Resources: Version 1*. DWAF, Pretoria, South Africa.

Ellis, J.B., Deutch, J.-C., Mouchel, J.-M., Scholes, L. and Revitt, M.D. 2004. Multicriteria decision approaches to support sustainable drainage options for the treatment of highway and urban runoff. *Sci. of Tot. Env.*, Vol. 334–35, pp. 251–60.

Environment Canada Task Group 1995. The ecosystem approach: getting beyond the rhetoric. Unpublished report, Environment Canada, Ottawa, Sept. 26, 1995.

Horner, R.R., Skupien, J.J., Livingston, E.H. and Shaver, H.E., 1994. *Fundamentals of urban runoff management: technical and institutional issues*. Terrene Institute, Washington DC.

Horner, R., May, C., Livingston, E., Blaha, D., Scoggins, M., Tims, J. and Maxted, J., 2002. Structural and nonstructural BMPs for protecting streams. B.R Urbonas. (ed.) *Linking Stormwater BMP Designs and Performance to Receiving Water Impact Mitigation*, American Society of Civil Engineers, New York, NY, pp. 60–77.

Intergovernmental Panel on Climate Change (IPCC), 2007. *Climate Change 2007: The physical science basis. Summary for policymakers*. IPCC Secretariat, WMO, Geneva, Switzerland.

International Joint Commission (IJC), 1978. *Great Lakes Water Quality Agreement of 1978*. IJC, Windsor, Ont., Canada.

International Joint Commission (IJC), 1988. *Great Lakes Water Quality Agreement of 1978, as Amended by Protocol Signed November 18, 1987*. IJC, Windsor, Ont., Canada.

Lakhan, C.V., Trenhaile, A.S. and LaValle, P.D., 2000. Environmental protection efforts in a developing country: the case of Guyana. Electronic Green Journal, Issue 13, Dec..

Lawrence, A.I., Ellis, J.B., Marsalek, J., Urbonas B. and Phillips B.C. 1999. Total urban water cycle based management. *Proc. of the 8th Int. Conf. on Urban Storm Drainage*, Sydney (Australia), I. B. Joliffe, and J.E. Ball (eds), Aug. 30- Sept. 3, 1999, Vol. 3, pp. 1142–49.

Marsalek, J., 2005. The current state of sustainable urban stormwater management: an international perspective. *Proc. Int. Workshop on Rainwater and Reclaimed Water for Sustainable Water Use*, University of Tokyo, Tokyo, June 9–10, 2005, pp. 1–11.

Marsalek, J., Watt, W.E. and Anderson, B.C., 2006. Trace metal levels in sediments deposited in urban stormwater management facilities. *Wat. Sci. Tech.*, Vol. 53, No. 2, pp. 175–83.

McLaren, R.A. and Simonovic, S.P. 1999. Data needs for sustainable decision making. *Int. J. of Sustainable Development and World Ecology*, Vol. 6, pp. 103–13.

New South Wales Fisheries, 1994. New South Wales Fisheries Act 1994, Government of New South Wales (NSW), Fisheries, Cronulla NSW, Australia (available on the website: www.fisheries.nsw.gov.au).

The Ramsar Convention on Wetlands, 1998. *Strategic Approaches to freshwater management: background paper – The Ecosystem Approach*. Commission on Sustainable Development, 6th Session, New York, USA (available at: http://www.ramsar.org/key_csd6-iucnwwf_bkgd.htm).

Roesner, L.A., Bledsoe, B.P. and Rohrer, C.A., 2005. Physical effects of wet weather discharges on aquatic habitats – present knowledge and research needs. Erikson, E., Genc-Fuhrman, H., Vollertsen, J., Ledin, A., Hvitved-Jacobsen, T. and Mikkelsen, P.S. (Eds.), *Proc. 10th Int. Conf. on Urban Drainage* (CD-ROM), Copenhagen, Denmark, Aug. 21–26, 2005.

Sahely, H.R., Kennedy, C.A. and Adams, B.J. 2005. Developing sustainability criteria for urban infrastructure systems. *Can. J. Civ. Eng.*, Vol. 32, No. 1, pp. 72–85.

Schueler, T.R. 1987. *Controlling Urban Runoff: A Practical Manual for Planning and Designing Urban BMPs*. Washington Metropolitan Water Resources Planning Board, Washington DC.

Strecker, E. 2001. Low-Impact Development (LID) – Is It Really Low or Just Lower? *Linking Stormwater BMP designs and performance to receiving water impacts mitigation*. B.R. Urbonas (ed.) Proc. of UEF conference. Snowmass, CO (USA), published by ASCE, Reston, VA. ISBN 0-7844-0602-2, 210–222.

Sverdrup-Jensen, S. 1997. Policy issues deriving from the impact of fisheries on food security and the environment in developing countries. Proc. International Consultation on Fisheries Policy Research in Developing Countries: Issue, priorities and needs. Disc. Paper 4, The North Sea Centre, Hirtshals, Denmark, June 2–5.

UN Division for Sustainable Development, 2005. *Plan of implementation of the World Summit on Sustainable Development*. UN, New York, US.

UN Division for Sustainable Development, 2006. *Revising Indicators of sustainable development – status and options*. UN, New York, US.

Urbonas, B. and Jones, J.E. 2002. Summary of emergent urban stormwater themes. B.R. Urbonas, (ed.) *Linking Stormwater BMP Designs and Performance to Receiving Water Impact Mitigation*, American Society of Civil Engineers, New York, NY, pp. 1–8.

Wolf, G. and Gleick, P.H. 2003. The soft path for water. P.H. Gleick (ed.) *The World Water 2002–2003*, Island Press. Washington DC, US, 1–32.

Wrona, F.J. and Cash, K.J. 1996. The ecosystem approach to environmental assessment: Moving from theory to practice. *Journal of Aquatic Ecosystem Health*, Vol. 5, pp. 79–87.

Yale Center for Environmental Law and Policy (Yale University) and the Center for International Earth Science Information Network (Columbia University). 2005. *2005 Environmental Sustainability Index*. Produced in collaboration with World Economic Forum, Geneva, Switzerland and Joint Research Centre of The European Commission, Ispra, Italy. Yale University, New Haven, Conn., US.

Zalewski, M. 2000. Ecohydrology – the scientific background to use ecosystem properties as management tools towards sustainability of water resources (guest editorial). *Ecological Engineering*, Vol. 16, pp. 1–8.

Zalewski, M., Wagner-Lotkowska, I., and Robarts, R.D. (eds) 2004. *Integrated watershed management – ecohydrology and phytotechnology – manual*. ISBN 92-9220-011-9, ISBN 83-908410-8-8, UNESCO, Paris.

Chapter 4

Ecosensitive approaches to managing urban aquatic habitats and their integration with urban infrastructure

Jiri MARSALEK[1], Diederik ROUSSEAU[2], Peter VAN DER STEEN[2], Sophie BOURGUÈS[3], Matt FRANCEY[3]

[1] National Water Research Institute, Environment Canada, Burlington ON, Canada L7R 4A6
[2] UNESCO-IHE (Institute for Water Education), PO Box 3015, 2601 DA Delft, The Netherlands
[3] Melbourne Water, PO Box 4342, Melbourne 3001, Victoria, Australia

The contributions of Diederik Rousseau and Peter van der Steen were supported by funding from the EU-SWITCH project (FP6, contract no. 018530).

The first part of this book provides the reader with basic information on aquatic habitats and their functioning. It reviews the existing strategies for urban aquatic habitat protection and rehabilitation, together with examples of related policies and regulations. Building on this knowledge, the second part of the book provides practical examples of technical measures applied to urban aquatic habitat management. First, this chapter provides examples of eco-sensitive approaches meeting multiple, sometimes conflicting, goals, where aquatic habitats may serve the urban population needs without losing their own ecological values and functions. Specific issues address interactions of UAHs with the city infrastructure, including water supply systems, urban drainage and flood protection, and wastewater management and sanitation. Some other scientific and technical principles and measures dealing with habitat rehabilitation, eco-hydrology and urban planning are presented in Chapters 5, 6 and 7, respectively.

4.1 URBAN WATER CYCLE AND AQUATIC HABITATS

Whenever the natural sources and pathways of water in urban areas no longer suffice to support and protect the population served, man-made infrastructures are built for providing water supply, urban drainage and flood protection, collection and management of wastewater, and the conservation or protection of ecosystems in the affected areas. Historically, such infrastructures have evolved as central systems designed to provide the required capacities, with minimal considerations of impacts on the environment and aquatic ecosystems. This approach is also referred to as the 'hard path' for water, reflecting the use of hard construction materials (like concrete) for building centralized infrastructures. However, this philosophy is rapidly changing, as new strategies promote sustainable systems based on the conservation and preservation of the environment and natural resources. In water management, this new strategy is represented by the 'soft path' for water, which promotes non-structural, policy-oriented strategies that focus on problem prevention and the use of both centralized and decentralized

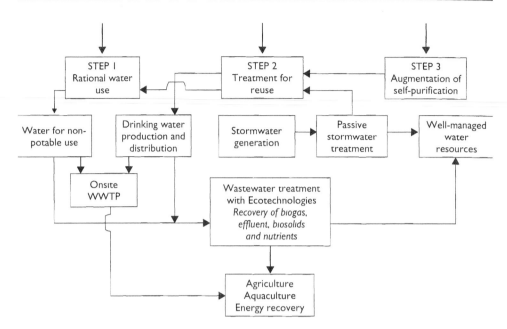

Figure 4.1 A three-step strategic approach to urban water management

Source: adapted from Nhapi and Gijzen. 2005.

facilities, as discussed in Chapter 3. The three major shortcomings of conventional water management avoided by using the 'soft path' include the following:

- Irrational water use: using water of high quality for purposes that can tolerate a lower grade
- Excessive costs: Gijzen (1997) estimated that some developing countries investing 1.5% of their Gross National Product (GNP) in environmental protection would require implementation periods exceeding the economic life span of the required facilities
- Unbalanced nutrient cycles: costly removal and disposal of nitrogen, rather than its reuse as a fertilizer, is inefficient in terms of both energy and resource utilization.

To overcome the above-mentioned problems, Nhapi and Gijzen (2005) developed the three–step strategic approach (see Figure 4.1), which specifically aims to change the urban water cycle and reduce the overall negative environmental impact of water use and thereby increase the sustainability of the system by the following means:

- Rational water use: applying water-saving measures and using lower quality water for non-potable purposes
- Treatment for reuse: wastewater is considered a resource
- Augmentation of natural self-purification, where wastewater discharge is unavoidable.

The subsequent sections will deal with habitat interactions among water supply, urban drainage and flood protection, wastewater management and sanitation. The issues of habitat restoration and the augmentation of self-purification capacity are of such a high importance that they are covered separately in Chapter 5.

4.2 HABITAT INTERACTIONS WITH WATER SUPPLY

Water supply withdrawals from urban waters (aquatic habitats) appear to have a limited effect on habitat functions, mainly because of relatively limited rates of withdrawal. However, a better appreciation of the effects of water supply becomes evident when examining the urban water cycle in its entirety, indicating that water supply represents a major importation of water into urban areas, and a high percentage of such water will re-enter urban waters as treated wastewater. Thus, the two main interactions between urban water supply and aquatic habitats can be identified as those caused by (a) the quantity of water imported into urban areas, and to a large extent discharged back into the environment as fully or partly treated wastewater, and (b) changes imposed by water withdrawal and distribution structures. The ecological implications of these changes will depend on the affected ecosystems themselves – in urban areas where riverine ecosystems are naturally seasonal, anthropogenic discharges that result in a change to perennial systems can have strong impacts on aquatic habitats (Day et al. 2005). In the soft path for water, solutions can be found by reducing demands for water, using water of the quality adequate for the particular demand, water re-use and by preventing or reducing pollutant loads being discharged via the drainage system. This is in agreement with the three-step system proposed by Nhapi and Gijzen (2005).

4.2.1 Managing water import into urban areas

Municipal water use includes all water delivered by the municipal water supply system. Actual per capita water usage ranges from 5 to more than 500 L per day (L/d), with the Basic Water Requirement (BWR) recommended as 50 L/d. In fact, more than 1 billion people use less than the BWR. Consequentially, there are challenges at both ends of the water-use scale: improving the supply of water at the lower end of the scale and implementing conservation measures at the upper end. For demand-side management, various instruments have been developed for different sectors and can be classified as water conservation measures, economic measures, information and educational measures, and legal measures.

The need for water conservation is accepted in most jurisdictions as the best solution for meeting the future water demands. Towards this end, management measures focus on reducing water losses and unaccounted for water, including the following:

- leakage from pipes, valves, meters, etc.
- leakage and losses from reservoirs, including evaporation and overflows
- water used in the treatment process (back-wash, cooling, pumping, etc.) or for flushing pipes and reservoirs.

Typically, these losses may vary from 10 to 60% (Marsalek et al., 2007), with the upper values reported for developing countries. Other measures, and particularly those promoted under the soft path for water approach, focus on an improved efficiency of water use, including water saving technologies, such as dual-flush toilets, flow restrictors on showers and automatic flush controllers for public urinals, automatic timers on fixed garden sprinklers, and moisture sensors controlling irrigation in public gardens. Full-cost pricing also strongly influences the water use.

The other important soft path consideration is matching the water quality supplied with the intended use requirements. While this approach does not necessarily reduce water use, it can save high quality drinking water and facilitate water reuse/recycling, thus avoiding water withdrawal from natural water bodies. Two such supplementary sources of water are particularly important (in terms of volumes) in urban areas: rainwater harvesting and wastewater reclamation and reuse.

Rainwater harvesting can be a supplementary or primary water source for individual households or small communities, especially in locations with seasonal rainfall or limited surface water resources (e.g., small islands). The use of roof runoff or good quality stormwater for landscape irrigation or industrial uses (e.g., concrete production) in highly developed urban areas, or low flow maintenance in urban streams (see also Chapter 5) is attracting interest as a measure supporting environmental sustainability by reducing water supply demands and reducing urban runoff flows and their impacts.

Rainwater and stormwater reuse contributes to the protection of urban aquatic habitats by preserving flow regime in three ways: reducing withdrawals from urban water bodies; reducing wastewater effluent discharge; and avoiding high stormwater peak flows, thus protecting the habitats and biotic structures supported by them from physical destruction.

The design of rainwater harvesting systems requires considerations of both water quantity and quality. With respect to water quantity, the important factors are local climate (particularly the annual precipitation and its temporal distribution), the rainwater collector area and storage. The quality of collected rainwater may match that required by the intended use; if not, some form of treatment may have to be implemented. The most feasible reuse of rainwater in urban areas is for irrigation, which may account for 35 to 50% of domestic water use in Canada and the US; other uses may be restricted by local regulations. Cowden et al. (2006) estimate that domestic rainwater harvesting could improve the per capita access of African urban slum dwellers to water, in the range of 20 to 50 L/d during at least three months of the year. A major limitation of rainwater reuse is the limited reliability of supply (depending on rainfall occurrence), which can be improved by adding storage to the reuse system.

Wastewater reclamation and reuse have been practised in many parts of the world for thousands of years. Steady supply is the main advantage of reclaiming and reusing wastewater, usually for such reasons as water scarcity or environmental considerations. A second advantage is the efficient use of wastewater nutrients as a substitution for chemical fertilizers. Indeed, demonstration fields have shown that the yields of crops irrigated with wastewater is comparable to that achieved by chemical fertilizers. Typical reclaimed wastewater uses include unrestricted urban and recreational uses and agricultural irrigation of food crops; restricted-access urban use includes restricted recreational use and agricultural irrigation of non-food crops processed before consumption, and industrial reuse and recycling. In one case, the city of Windhoek, Namibia, reclaimed wastewater has been used to supplement potable water supply since 1968 (Metcalf and Eddy, 2003). In other cases, low-polluted wastewater or grey water is even used in aquaculture. Wastewater reuse is most common in Israel, where in some locations, 90% of wastewater is reused.

Well-known operating wastewater reuse schemes include NEWater in Singapore (1% of drinking water consumption is covered with reclaimed wastewater), where the quality of the reclaimed water fulfils all requirements and is in most aspects better than that of the raw source water used otherwise. In the Shinjuku district of Tokyo, the water recycling centre distributes treated wastewater for toilet flushing in twenty-six high-rise office buildings. In Mexico City, untreated sewage has been used to irrigate the Mezquital Valley since 1896. As a result of this practice, the water table of the aquifer underlying the irrigation zone has been rising. The unplanned artificial recharge is about 25 cubic metres per second (m^3/s), and this 'reclaimed water', treated only with chlorine, is being used to supply 300,000 inhabitants of the region for human consumption. Several studies have shown that the water meets potable norms and WHO human consumption guidelines, including toxicological tests (Jiménez et al., 2001).

Sydney Water (in Australia) encourages the reuse of stormwater and grey water in addition to ongoing intensive water saving and conservation campaigns. The use of reclaimed water for garden irrigation is encouraged, using treated grey water or (untreated) stormwater collected in separate tanks. Sydney Water also introduced the distribution of reclaimed water in dual water supply systems, providing reclaimed water and drinking water. To ensure that the drinking water is not confused with the recycled water, it is delivered by a separate distribution system. The reclaimed water is subject to strict guidelines that limit its use to toilet flushing and outdoor purposes, such as car washing and garden irrigation (NSW Health Department, 2005). Such measures can be incorporated in new developments more economically than in existing ones in which the infrastructure has already been installed.

A recent survey of more than 3,300 water reclamation projects revealed that technological risks no longer represent a major concern for development of water reclamation; more critical issues seem to be social acceptance, financing and failure management (Bixio et al., 2005).

Finally, protecting receiving aquatic habitats not only means reducing the volumes of wastewater, but also the pollutant loads. A well-known example is the removal of phosphorus from laundry detergents in the mid 1970s in the US. This action, together with reducing pollutant loads of organic matter, led to a great improvement of water quality in Lake Erie, which was declared 'dead' during the 1960s (Sweeney, 1995). Similar actions are reported elsewhere by Nhapi and Gijzen (2005). Tangsubkul et al. (2005) used substance flow analysis to assess imports and exports of phosphorus (P) in Sydney, Australia for the year 2000 and estimated that 80% of the import occurred via food and detergents, and as much as 25% of the imported P-mass eventually ended up in the ocean, even after wastewater treatment.

4.2.2 Impacts on aquatic habitats imposed by water reservoirs

Conjunctive use of surface water and groundwater in water supply is common in many regions. In some countries, however, the use of surface water dominates, providing water for 60% (USA), 65% (UK), 75% (Canada), 80% (Spain), or even 85% (Norway) of the total population (Marsalek et al., 2006). While lakes represent a relatively steady source of water, rivers require flow regulation achieved by constructing dams and creating water reservoirs. In most cases, such reservoirs are built as multi-purpose facilities, which in addition to water supply, may also provide flood protection, agricultural irrigation, power generation, aquaculture, fishery, recreation, and other functions.

BOX 4.1 Polderdrift Case study, Arnhem, The Netherlands

In Arnhem, grey water from 40 houses in the Polderdrift community is treated using a subsurface flow constructed wetland and reused for toilet flushing. Rainwater is harvested and used partly for washing laundry and partly infiltrated in-situ. The black water is conveyed to a conventional wastewater treatment plant. Van Betuw (2005) reported that this scheme resulted in lowering water consumption by 57% and reducing wastewater discharge by 85%. It is of interest to note that the tenants were involved during the design phase in the selection of environmental measures and continue their involvement by conducting simple maintenance tasks. Besides the environmental benefits, there is also a financial incentive for tenants who pay less for water and wastewater services. There is a continuous need to advise new tenants about the nature of their water distribution system to avoid cross-connections between the potable water and low-quality water distribution networks.

Other successful applications of grey water and rainwater use in urban areas have been reported by Jeppesen (1996), Nolde (1999), Friedler and Hadari (2006), Goddard (2006) and Ghisi and Ferreira (2007). Most authors agree that public acceptance is high, but also warn that the public has a wrong perception that grey water is safe. In order to safeguard public health, a continual process of verification, plumbing management and customer education is therefore needed.

The damming of rivers and the creation of reservoirs impact the river continuum and thus the integrity of major processes in a catchment, including the water cycle dynamics and downstream transport of matter (Tarczynska et al., 2002). This also represents a major change of the aquatic habitats features (see Chapter 2) over long reaches extending upstream and downstream of the dam and impacts the structure and functioning of associated biotic communities.

Physical habitat changes start with increased depths, water levels and shifts of the aquatic ecosystems boundaries and are usually accompanied by losses of shoreline vegetation and wetlands. Shoreline slumping and erosion provide a source of sediment depositing elsewhere and degrade micro- and macro-habitat features, thus reducing refuges and spawning grounds for fish.

Physical habitat changes also result from modifications of flow patterns, including changes in flow velocity, flood timing and magnitude, and increasing hydraulic residence time, among others. These further modify the structure of species and their domination and may, for example, favour production of phytoplankton over macrophytes. Modification of seasonal flow patterns and water-level variations may interfere, for example, with instream and ecotone vegetation succession and fish life cycles and lead to a loss of refuge for zooplankton and fish species. Some organisms (zooplankton, phytoplankton and fish) may be seasonally impacted by being swept away from reservoirs by high flows and generally suppressed fish migration.

Water quality changes resulting from damming include the increased retention of matter transported in from the upstream catchment and modification of nutrient dynamics and energy flow within the impounded stretch of river. Such changes manifest themselves by increased concentrations of nutrients, humic substances and organic matter, and in urbanized catchments, by increased concentrations of heavy metals, pharmaceuticals, personal care products and trace organic substances that may affect human health (see Chapter 8). The increased retention of nutrients often causes secondary pollution, resulting from over-productive species of (often toxic) cyanobacteria and may result in the exclusion of an impoundment from drinking water supply and recreational use. Water quality changes also include decreases in dissolved oxygen concentration due to the decomposition of biodegradable organics and changes of thermal regime.

These disturbances may lead to the loss of species and thereby affect biotic inter-actions in ecosystems and reduce food (energy) sources for some aquatic biota. Changes in the trophic structure transfer across the trophic pyramids and intensify the effects at other trophic levels (the 'top-down' effect). For example, the loss of shallow areas, marshes and wetlands may reduce the diversity and abundance of some insects that naturally constitute food for fish, thereby reducing fish populations and fishery yields.

These examples show that the disturbance of the river continuum modifies several environmental processes, such as ecological succession, biodiversity sustenance and the biological productivity of aquatic and adjacent terrestrial habitats. Both free-flowing and impounded systems support different organisms (ranging from phyto- and zoo-plankton, through different communities of benthic organisms to fish) and trophic rela-tions among them and also support other types of wildlife, including amphibians, reptiles, waterfowl, predatory birds, and mammals.

In summary, when planning water supply in urban areas, and particularly where sup-ply rates are growing by necessity, one should always recognize, and account for, the resulting impacts on the aquatic habitats and the related ecological amenities. These impacts include the importation of large quantities of water, which largely changes into wastewater (discussed later in more detail) as well as the need for river damming to ensure water withdrawals and the associated plethora of aquatic habitat changes, includ-ing the loss and fragmentation of habitat. Such negative impacts may be mitigated to a certain extent by applying ecosensitive measures, which are complementary to hydrotech-nical structures, as discussed elsewhere in this chapter and in Chapters 5 and 6.

4.3 HABITAT INTERACTIONS WITH URBAN DRAINAGE AND FLOOD PROTECTION

The protection of urban areas against flooding and provision of convenience by pre-venting water ponding and the disruption of activities in urban areas are among the basic requirements of urban water services. Urban floods can be either locally gener-ated by high intensity rainfall (also referred to as pluvial floods) or generated in larger river catchments and pass through urban areas, where they may inundate built up areas (also referred to as fluvial floods). Only the first class of floods, locally gener-ated, will be addressed here; the second group is beyond the scope of this book, as are the other types of floods that may occur in coastal areas in the form of storm surges or tsunamis.

4.3.1 Urban drainage: Problem definition and needs for management

4.3.1.1 Urban drainage impacts on aquatic habitats

Urban drainage exerts strong effects on aquatic habitats in urban areas in two ways: first, through impacts on the existing habitat, and second, by creating new habitats in the form of urban stormwater facilities, including stormwater ponds and wetlands.

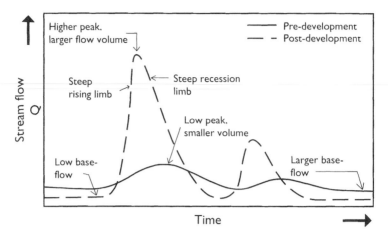

Figure 4.2 The effects of urbanization on streamflow
Source: adapted from Schueler, 1992.

Urbanization exerts profound effects on the hydrologic cycle of the affected areas. The understanding of such effects has evolved gradually over the past forty years, starting with Leopold's (1968) seminal study addressing the water balance in urbanizing areas and explaining the effects of increased imperviousness on reduced evapotranspiration and infiltration, higher surface runoff and reduced groundwater recharge. Concerns about increased runoff and the resulting flooding then led to extensive research on runoff peak increases due to urbanization, with a large number of publications assessing (mostly by modelling) changes in runoff peaks from urbanizing watersheds depending on the degree of development (imperviousness) and the return period of the rainfall/runoff event (Roesner et al., 2005).

Urbanization affects surface runoff by increasing runoff volumes, due to reduced rainwater infiltration and evapotranspiration; increasing the speed of runoff, due to hydraulic improvements of conveyance channels; and reducing the catchment response time and thereby increasing the maximum rainfall intensity causing the peak discharge. Urbanization thus changes the catchment hydrologic regimen. Efforts have been made to quantify these changes in the literature, with Riordan et al. (1978) estimating increases in the mean annual flood ranging from 1.8 to 8 times pre-urbanization levels, and the 100-year flood increasing from 1.8 to 3.8 times due to urbanization. The volume of direct stormwater runoff increased for various return periods up to six-fold. In general, the magnitude of such increases depends on the frequency of storms, local climate and catchment physiographic conditions (soils, degree of imperviousness, etc.). Figure 4.2. illustrates the differences in stream hydrology before and after urban development.

Next to the hydrological impact, urban runoff also affects aquatic habitats by conveying a broad range of pollutants originating from different sources, including accumulations on impervious surfaces during dry weather. A summary of stormwater quality data from several databases is presented by Marsalek et al. (2007). Among various sources of stormwater runoff, highway runoff is particularly polluted and represents a significant source of suspended solids, heavy metals, hydrocarbons and de-icing salts discharged to receiving waters (Sriyaraj and Shutes, 2001).

4.3.1.2 Overview of aquatic life support conditions in urban waters impacted by stormwater discharges

The major factors affecting the performance of biological communities have been introduced in Chapter 2. Two of them, flow regime and habitat structure, are particularly affected while considering the interaction between UAH and stormwater discharges.

Flow regime: Because of its crucial importance for habitat functioning, current methods of stormwater management strive to mitigate changes of flow regime, with various degrees of success, by attempting to preserve the pre-development water balance (Jones et al., 1997; Roesner et al., 2005; Urbonas and Jones, 2002). However, in many cases, mitigation against fundamental levels of change, such as transition from a naturally seasonal to a perennial system, may not be realistic in an urban context. The ecological implications of this depend on the type of ecosystem that is affected – in Cape Town, South Africa (see Chapter 9), an urban area in the heart of the Cape Floral Kingdom, where endemism is high and where the natural condition of many of the river and wetland systems would be seasonal, the biodiversity implications of such changes may be profound (Day et al., 2005).

Discharges of stormwater runoff and effluent into naturally seasonal rivers may also have profound ecological effects, even when water quality impacts are discounted. Perennial flows in a naturally seasonal river can dilute seasonal cues, important for the life history patterns of many organisms; encourage the proliferation of invasive, often alien aquatic plants and fish species at the expense of seasonally variable communities that would have occurred naturally in these systems; and alter surface-groundwater interactions, with implications for other interlinked systems, such as seasonal wetlands. In the Cape Floral Kingdom of Cape Town, South Africa, for example, up to 97% of naturally seasonal wetlands, many of which were linked to seasonal river systems, have been permanently altered as a result of increases in the water table and a loss of seasonality, primarily as a result of effluent discharges (Day, 1987). Since this region has high levels of endemicity, the implications of these changes for biodiversity, on a global scale, may be high.

The physical containment of flows within urban areas and the need to control velocity and flood volumes contribute to the low levels of effective ecological sustainability of most urban aquatic ecosystems, in which the absence of natural channel maintenance and wetland rejuvenation processes (e.g., scour) are absent, making ongoing maintenance of such systems a reality (Roesner et al., 2005).

Habitat structure: In general, a complex and variable physical habitat contributes to good performance of fish and benthic communities (see Chapter 2). However, stormwater impacted habitats are characterized by simplified habitat structures and reduced biological community performance (Roesner et al., 2005). Studies of natural fish habitats (for various salmon species and life stages) have identified required conditions with respect to minimum depths, velocity range, substrate size range, temperatures, minimum dissolved oxygen, and minimum space (i.e., area per fish). In typical stormwater impacted habitats, such conditions are not met (Horner et al., 1994).

4.3.1.3 Need for stormwater management

In this book, the discussion of urban drainage issues focuses on drainage impacts on aquatic habitats. It should be recognized, however, that the reasons for addressing the problems caused by urban runoff vary in individual countries and regions and shape the stormwater management approaches taken. This has been confirmed by the International Water Association (IWA), whose international report on stormwater management (Marsalek and Chocat, 2002) indicates a widespread acceptance of a holistic approach to stormwater management promoting sustainable urban drainage systems, with an emphasis on stormwater reuse, source controls, transition from traditional 'hard' infrastructures to green infrastructures, needs for infrastructure maintenance and rehabilitation, formation of stormwater management utilities, sustainable funding through drainage fees, and active public participation in the planning and operation of such systems. While in the countries with predominantly separate sewerage (e.g., Australia, Canada, USA) the concerns about the ecological health of urban streams and rivers are particularly strong (Walsh et al., 2004), in many European countries with combined sewerage systems, the prevention of recurrent flooding problems may be of greater importance. Wet weather events can have disastrous consequences within these impervious urban centres, and chronic combined sewer overflows are common. Such problems, among others, are explored by the case studies of the Lyon, France and Lodz, Poland, presented in Chapter 9. Stormwater management has to undergo a change with respect to ensuring public safety and protecting the environment.

Another driver for improved stormwater management is the potential use of stormwater as an alternative supply of water. Where stormwater harvesting is a viable option, positive water quality gains will have to be assessed against negative impacts on the receiving water bodies, which are often flow-stressed in urban environments. Costs and benefits, and above all public acceptance (Exall, 2002), also need to be considered when deciding on the end use chosen for the harvested water (Alluvium, 2006).

4.3.2 Urban drainage: Stormwater management goals

During the last fifteen to twenty years, stormwater management goals have expanded from the mitigation of urbanization impacts to a more proactive role, including the provision of environmental benefits or amenities, such as the creation of aquatic and wildlife habitat, preservation of the existing habitat and biodiversity (Horner et al., 2002), and the provision of community amenities (Urbonas and Jones, 2002). Consequently, more attention has been paid to mitigating the effects of urbanization on physical habitats in receiving waters, but with limited success as shown in a recent review paper (Roesner et al., 2005). Major findings of this review were as follows:

- Linkages between larger-scale processes (e.g., land cover changes, changes in catchment hydrology and runoff formation) and the small-scale habitat descriptors (e.g., pool-riffles structures, substrate rugosity) are largeley undocumented and poorly understood.
- The interpretation of biomonitoring data is done at two disparate spatial scales: ecoregions and local habitats.
- Flow regime, which has been identified as the key factor in formation of the physical habitat, is typically not measured (except for short periods of stream surveys).
- Limited empirical evidence indicates that stream habitats are simplified by the wet-weather flow regime modified by urbanization (Roesner et al., 2005).

Furthermore, it should be also recognized that within the time scales of geomorphic processes (thirty to fifty years), other driving forces also change, including catchment development/redevelopment, climate, and public expectations of environmental quality.

Besides stormwater effects on existing habitats, a related issue of interest deals with the creation of aquatic habitat in stormwater management facilities. Urban stormwater Best Management Practices (BMPs) were originally developed for mitigating the impacts of urbanization on the hydrological cycle and water quality, but such objectives were later expanded to include environmental amenities, such as the protection or creation of aquatic habitats (Jones et al., 1997). The issue of habitat creation is not without some controversy, because these new demands on BMPs may be in conflict with the fact that BMPs are primarily passive treatment facilities designed to retain sediment and pollutants. The issues of habitat conditions in stormwater facilities were addressed by Bishop et al. (2000a, 2000b) who focused on water quality and toxicity issues at fifteen stormwater management facilities, including eleven stormwater ponds and four constructed wetlands. The study found that wildlife used all of the fifteen stormwater facilities, with species richness classified as low to moderate; all stormwater facilities contained contaminants, generally at low levels, except for some persistent contaminants in sediment and water exceeding the Canadian guidelines for water and sediment quality in the freshwater environment; bioaccumulation of some persistent contaminants (e.g., PCBs) in red-winged blackbird eggs was found at two sites; and toxicity findings – sediment from one site in a commercial/light industrial area was toxic to invertebrates in a short-term bioassay, and conditions were toxic to frog development at one of four residential sites studied, and no site contained sediments toxic to fish in short term bioassays. Consequently, the authors concluded that the stormwater facilities studied (built in the mid-1980s) did not provide good habitats for fish and wildlife, because of their potential for contamination.

There is also an argument to be made for the separation of functions in different structures or sections of aquatic systems. For example, *Typha capensis* reedbed marshes are associated with relatively low habitat quality but high levels of ecosystem services, in terms of water quality amelioration. This service will be diluted by attempts to increase habitat diversity in such areas. Where water quality is a limiting factor on aquatic habitat quality, upstream areas may be beneficially sacrificed to improving water quality, to the benefit of areas further downstream, in which efforts to improve habitat diversity or quality may have better success. Similarly, ecologically sterile silt traps may be a more efficient way of trapping intense sediment than vegetated wetlands, and the habitat quality of the latter may be better served by the incorporation of such hard engineering structures, provided that the trade-off is the creation of less disturbed and/or more ecologically functional habitat downstream.

4.3.3 Tools for effective stormwater management

New approaches for dealing with the water cycle in urban landscapes have emerged since the mid-1970s[1] both in developed (Azzout et al., 1994; Schueler, 1987; Urbonas,

---

[1] For example, starting with the Woodlands master-planned community in Texas: www.thewoodlands.com.

Figure 4.3 An example of a stormwater pond in Toronto, Canada. Photo by Quintin Rochfort, Environment Canada (See also colour plate 1)

1994) and developing countries (Baptista et al., 2005; Parkinson and Mark, 2005). They adopt a holistic view and aim to integrate land-use and water managements. Generally known as BMPs, Water Sensitive Urban Design (WSUD), Sustainable Urban Drainage Systems (SUDS) or Low Impact Development (LID) techniques, presented earlier in Chapter 3 (US EPA, 2000), they provide a range of solutions aiming to achieve sustainable urban development that responds to hydrological, environmental and community concerns. An example of one such measure, a shallow stormwater pond/constructed wetland in Toronto, Canada, is shown in Figure 4.3. The overarching strategy is to mimic pre-development conditions of the hydrological regime, generally in lower development density parts of urban areas.

The concepts of BMP/WSUD/SUDS/LID are now increasingly applied to stormwater management and aim at reducing the impact of the volume, frequency, and quality of stormwater runoff entering waterways or other environmentally significant assets. This philosophy also foresees alleviating the pressure on potable water supply by encouraging the society to revaluate its behaviour towards water and by matching the use of water with its quality. At present, however, the experience with these new systems is limited, particularly in higher density developments, and it is uncertain that their performance will be sufficient to fully protect urban aquatic habitats, even while meeting the stormwater management guidelines and targets discussed in the next section.

4.3.3.1 Guidelines and targets

Numerous handbooks and design manuals, sometimes associated with computer models, are available to assist in the design and selection of BMPs. Examples include, the Washington, D.C. guide for BMPs (Schueler, 1987), the California Stormwater BMP Handbook (CDM et al., 1993), French BMP guides (Azzout et al., 1994; Bergue and Ruperd, 1994), the Ontario Stormwater Management Planning and Design Manual (MOE, 2003), UK CIRIA Reports C521 and C522 (CIRIA, 2000a; 2000b) and C523 (CIRIA, 2001), *WSUD Engineering Procedures: Stormwater* (Melbourne Water, 2005), *Australian Runoff Quality: A guide to water sensitive urban design* (Engineers Australia, 2005), and many others. A thorough review of technical aspects of the design for WSUD is provided in a report published through the EU 5th Framework Program (DayWater Project, 2003).

A clear objective of reducing targeted pollutants constitutes a tangible goal for planners and developers, but does not guarantee the protection of aquatic habitats by itself.

4.3.3.2 Technological measures (BMPs)

Typical technological measures for managing stormwater include infiltration facilities, ponds and wetlands, swales and ditches, oil and sediment separators, and real-time control operation systems. A brief overview follows, and details can be found in the above-listed manuals and references and in other books in this UNESCO series on urban water management (e.g., Marsalek et al., 2007).

Infiltration may be applied in percolation basins, trenches (designed as underground gravel-filled units), and wells. This technology has been used for a very long time, particularly on a small scale in rural settlements, but only during the last few decades has it been further developed and adopted in urban areas on a larger scale. Stormwater infiltration helps keep groundwater tables at a natural level, which promotes good conditions for vegetation and a good microclimate. The construction costs of drainage systems with infiltration facilities are also cheaper than those of conventional systems. Infiltration is also implemented on grass or other permeable surfaces and in drainage swales and ditches. The use of this measure is steadily growing in many countries.

In many countries, ponds and wetlands have become a common and accepted means of attenuating drainage flows and treating stormwater by removing suspended solids, heavy metals and, to some extent, nitrogen and phosphorus. The cost of construction and operation of such facilities is often low compared to their environmental benefits. The sediments from ponds may contain high concentrations of heavy metals; however, ponds and wetlands should be considered as stormwater treatment facilities and not as natural water bodies, even if they often provide aesthetic and habitat values to the urban area.

Swales and ditches are applied commonly in the upstream reaches of drainage systems to control runoff flows and provide runoff quality enhancement. Flow control is obtained by stormwater infiltration into the ground, quality enhancement by filtration through the turf, solids deposition in low flow areas, and possible filtration through a soil layer.

Oil and sediment (grit) separators are used to treat heavily polluted stormwater from highways or truck service areas, or where polluted stormwater is discharged into sensitive receiving waters. The efficiency of these units in trapping oil, sediments and chemicals attached to the sediments is often poor, because of under-sized units or lack of flow-limiting devices preventing the washout of trapped materials.

Real-time control operation of sewer systems has been developed during the last two decades and implemented in some Canadian, European, Japanese, and US cities. The applications are often in combined sewer systems, and the purpose is mainly to reduce combined sewer overflows and/or the overloading of wastewater treatment plants by the maximum utilization of the dynamic capacities of the system (Colas et al., 2004).

In some jurisdictions, drainage is provided via combined sewers, which convey both sanitary sewage and stormwater. The extent of combined sewers varies from country to country from 20 to 90%. Generally, combined sewers are more common in climates with lower annual rainfall; for high rainfalls, the system would be too overloaded and collect a low percentage of total flows.

Combined Sewage Overflows (CSOs) are caused by excessive inflows of stormwater into the sewer system, so any measure reducing stormwater runoff and its inflow into combined sewers would also help abate CSOs. The control of overflows is accomplished by various forms of flow storage and treatment; flow storage serves to balance CSO discharges, which may be returned to the treatment plant after the storm, when flows have subsided below the plant capacity (Marsalek et al., 1993).

CSO storage can be created in a number of ways: by maximizing the storage use available in the existing system – for example, through the centrally controlled operation of dynamic flow regulators in real time (Schilling, 1989) – as newly constructed storage online or off-line (online storage includes oversized pipes or tanks; off-line storage includes underground storage tanks or storage and conveyance tunnels), or even in receiving waters, as in the flow balancing systems created by suspending plastic curtains from floating pontoons, in a protected embayment in the receiving waters (WPCF, 1989).

4.3.4 Promotion of modern stormwater management

Stormwater management needs to be socially marketed to increase its uptake and adoption in individual jurisdictions. The important points in this process include demonstration projects, public awareness, education and participation, favourable institutional framework, and the strategic positioning of stormwater management programmes within the overall concept of sustainable development.

One of the keys in the adoption of WSUD in Melbourne was the completion of actual demonstration projects (Lloyd, 2004; Brown et al., 2005), demonstrating the fact that learning from other countries is necessary, but local examples are more powerful. Such projects crystallize the partnerships between government, developers and scientists. Fostering these interactions is essential for the acceptance and implementation of the WSUD philosophy. At present, training and education tools are greatly tailored to the needs of local planners and developers, but it is widely acknowledged that the successful integration of WSUDs should not be left solely to their willingness. Market mechanisms can facilitate the implementation of modern stormwater management, by such tools as rebates on rainwater tanks, stormwater management fee credits for implementation of BMPs, or offset fees when stormwater quality standards are not met (Lloyd et al., 2004; RossRakesh et al., 2006).

Public awareness, education and participation are essential for successfully planning, implementing and accepting modern stormwater management approaches, promoted under such names as WSUD, SUDS, or LID. To this end, responsible agencies need to increase public recognition and understanding of drainage problems and their remediation, including the identification of responsible parties and past efforts, acceptance of the community ownership of these problems and solutions, and the need for moves towards environmentally responsible practices and behaviour of both individuals and corporations, with the integration of public feedback into programme implementation (WEF and ASCE, 1998; Victorian Stormwater Committee, 1999; UNEP, 2000). The success of modern stormwater control measures involving on-site facilities depends on public education and participation. While the benefits of stormwater management practices are recognized and relatively easy to demonstrate for structural facilities, it is much more difficult to assess the effectiveness of distributed source controls and the related Public Awareness/Education/Participation Programmes (PAEPP). Such programmes can be developed using the following steps:

- Define and analyse the problem (the sources of pollution)
- Identify stakeholders (commercial business, industry, land holders and residents, school/youth groups, and municipal staff)
- Know the target group: establish a complete group profile, develop the best methods of communication
- Set objectives: informative messages, emotional messages, responsibility messages, empowering messages, action messages (clear, simple language; technically sound statements; break up the concept into simple statements)
- Design methods by selecting techniques suitable for the group targeted
- Form action plans and timelines: identify costs, funding sources and trim the project to fit resources
- Monitor and evaluate: collect information and records to see how effective it is, recognizing that there will be lag in public response (Victorian Stormwater Committee, 1999).

Common elements of PAEPP's include printed materials; media; signs; community programmes; displays; community water quality monitoring programmes; official activity launches; local action committees and groups; advisory groups; and consumer, business and school education programmes. Many municipalities in the US found storm drainage system signs with prohibitive language and icons discouraging polluting to be particularly effective (WEF and ASCE, 1998). To reduce costs, this activity can be implemented mostly with a volunteer workforce, but in high traffic zones, municipal staff should be employed for safety reasons.

Public awareness and education is implemented through public meetings, open houses, tours of facilities, and visual displays at stormwater management sites. In this process, concerned citizen and environmentalist groups are formed, which then actively engage in environmental projects, including organized cleanup and publicity campaigns, and help involve schools (WEF and ASCE, 1998; Victorian Stormwater Committee, 1999).

Without public awareness and education, stormwater may be difficult to manage, poorly understood and not always perceived as a threat to the receiving waters. In some countries, such as Australia, the water pollution issue may not appear as pressing as climate change (Research Wise, 2006), perhaps because of strong concerns about extreme drought conditions. An innovative way of raising public awareness about the gross pollution of urban waters has been demonstrated in Melbourne, Australia by dumping tagged litter in Melbourne and recovering it (up to 95%) in the catchment waterways (McKay and Marshall, 1993). However, comparable methods are not available for more subtle pollutants.

Dry climate seems to be a likely reason why the adoption of WSUD is less developed in Southern European countries (Spain, Italy, Greece and Portugal) as compared to those in Northern Europe (DayWater Project, 2003). In such climatic conditions, one should emphasize other benefits of stormwater management, including stormwater reuse by harvesting and alleviating both the pollution and water shortages.

Much has been written about the need for a good institutional framework for effective environmental management, including the implementation of sustainability (see Chapter 3). The same concern applies to stormwater management, whose implementation has been impeded in the past by fragmented jurisdictions and numerous stakeholders with sometimes conflicting interests (Alluvium, 2006). Effective institutional frameworks are still evolving, without much consistency among various countries, or even within countries. The establishment of clear objectives is a key to enforceability. In most urban locations, there are various constraints on the adoption or implementation of BMPs, as discussed earlier.

The implementation of BMPs will only be successful if it contributes to attaining sustainability in urban areas, and if such contributions can be measured and assessed against sustainability indicators. Extensive listings of sustainability indicators and criteria are presented in Tables 3.1 and 3.2. While the experience with BMP/WSUD performance is still rather limited, numerous studies demonstrated, in specific conditions, that the use of BMPs can reduce drainage costs by 18 to 50% over conventional systems (DayWater Project, 2003). Furthermore, some of these measures are well suited even for highly urbanized areas: permeable pavements, roof gardens, bioretention systems in streets with little or no reduction of parking space, and rainwater tanks.

4.3.5 Flood protection

Locally generated floods usually result from catchment urbanization characterized by low hydrologic abstractions (high catchment imperviousness), hydraulically efficient flow conveyance, and reduced concentration times, generating high rates of runoff by high-intensity rainfall. The resulting runoff is conveyed either by storm sewers in separate sewer systems or by combined sewers in combined sewer systems, as discussed above.

Flood protection has been practised as a single-purpose water management measure for thousands of years, but increasing competition for water resources and conflicts among their uses has led to more holistic approaches in recent years. In the ecosystem approach, the flood management planning in catchments should be connected with land use planning (City of Cape Town, 2003) and should address not only flood defence, but also the protection of water resources, including fish habitats. Among the three possible approaches to flood management measures, i.e., living with floods, non-structural measures, and structural measures, only the third category exerts strong impacts on aquatic habitats. Structural measures reduce flood volumes and peaks by spatially distributed management measures serving to reduce runoff generation and structural measures including storage facilities, the enhancement of river bed flow capacity, and earthen platforms and polders in flood plains. All of these measures are designed to reduce flood volumes and peaks by storage and flow redistribution. Perhaps the largest impacts on habitat are caused by reservoirs, including the loss of land to water storage, loss of habitat, reduced biodiversity, interference with fish passage, the introduction of nuisance species and diseases, loss of land regeneration in flood plains, and the creation of new risks associated with potential dam failures.

In developed countries, mitigating urbanization impacts on runoff has been addressed with various degrees of success by applying both non-structural and structural control measures incorporated in a master drainage plan. The methodology for urban drainage planning is well developed and described by Geiger et al. (1987). A master drainage plan represents a technical layout of the sewerage systems (drainage and sanitation) for the entire urban area, as it may further develop within the planning horizon. With respect to drainage, the master plan should be a part of the catchment plan and incorporate the whole drainage system, including the connections between minor and major system components.

The situation is quite different in developing countries, where urbanization occurs too fast and unpredictably, and often progresses from downstream to upstream areas, which increases flood problems (Dunne, 1986). The urbanization of peri-urban areas is largely unregulated, often without the provision of any infrastructure, many public lands are occupied and developed illegally, and flood-risk areas (flood plains) are occupied by low-income populations without any protection. The World Health Organization has reported spontaneous housing developments in flood-prone areas of many cities in humid tropics, including Bangkok, Mumbai, Guayaquil, Lagos, Monrovia, Port Moresby and Recife (WHO, 1988).

Other problems include a lack of funding for drainage and other services, a lack of solid waste collection (which may end up in and block drainage ditches), no prevention of occupation of flood-prone areas, a lack of knowledge on coping with floods, and a lack of institutions in charge of flood protection and drainage (Dunne, 1986; Ruiter, 1990). Tucci and Villanueva (2004) suggest solutions like introducing better drainage policies, which would control flow volumes and peaks, and planned development, in which space is retained for flow management measures. Moreover, in flood plains, non-structural measures should be applied by emphasizing green areas, paying for relocation from flood-prone areas, and providing public education about floods.

4.4 HABITAT INTERACTIONS WITH WASTEWATER MANAGEMENT AND SANITATION SYSTEMS

In spite of continuing progress in providing the global population with improved sanitation, at the beginning of 2000, 2.4 billion people lacked access to improved sanitation (WHO and UNICEF, 2000; UNEP, 2003), with the majority of these people living in Asia and Africa. Further improvements are desperately needed. Wastewater generated in urban areas, and solid wastes, affect aquatic habitats, mostly with respect to water quality and habitat structure. Such impacts depend on the type of pollutant, the relative magnitude of the discharge in relation to the receiving waters, and the self-purification capacity of receiving waters.

4.4.1 Basic demands on wastewater management systems

Basic demands on modern urban wastewater management systems include the following requirements:

- Public health protection: must be capable of destroying or isolating faecal pathogens. This has been a major driving force behind the construction of wastewater treatment plants and introduction of effluent disinfection.
- Protect the environment: must prevent pollution of receiving waters and conserve valuable water resources. Typical wastewater pollution impacts include eutrophication, oxygen depletion, and toxicity.
- Recycle nutrients: must return nutrients in the wastewater to the soil. The nutrient contents of domestic wastewater may be valuable as fertilizer in agriculture (or aquaculture). However, the water treatment residue (sludge or biosolids) may be polluted with heavy metals and trace organic substances; measures must be taken to protect public health.
- Culturally acceptable: must be aesthetically inoffensive and consistent with cultural and social values; for example, in some cultures, dry sanitation is not acceptable.
- Reliable: must be easy to construct and robust enough to be easily maintained in a local context.
- Convenient: must meet the needs of all household members, considering gender, age and social status; carrying drinking water or wastewater products has strong gender implications and should be avoided in sustainable systems.
- Affordable: must be financially accessible to all households in the community; as a general guideline, the World Bank recommends that the cost of water and sanitation should not exceed 5% of the total family income.

Modern wastewater management systems meeting these demands are also referred to as ecological sanitation systems, which employ dry toilets, separating urine and faeces at the source (Winblad and Simpson-Hebert, 1998). The separated urine can then be transferred to the nutrient processing plants, where nitrogen and phosphorus are recovered and transformed into chemical fertilizers. If the urine and faeces are collected and used in agriculture, the remaining grey water is easier to treat by conventional wastewater treatment processes.

Wastewater management systems with separation sources have been operated, on a small scale, in a number of European countries. The advantages of such systems in rural areas of developing countries are striking: the hygienic conditions are improved compared to simpler solutions; water is only used for cleaning purposes; and wastewater products, urine and/or excreta, can be used as fertilizers after a minimum storage period.

The advantages of source separation systems in densely populated areas have not been proven in practice. Opponents argue that the cost for redesigning the sewerage systems in houses and streets will be huge, and that the transport of collected urine and/or excreta will cause additional costs, nuisance and air pollution in the cities. On the other hand, advocates argue that the cost/benefits of the system are favourable when compared to conventional systems, particularly after accounting for the recycling of nutrients and the use of natural resources. It would seem that the separation system may be feasible in peri-urban areas and informal settlements where sewerage systems still do not exist and the extension of the central system to the outlying parts of the city would be too costly or almost impossible due to the lack of space and financial resources.

4.4.2 Wastewater systems without separation of waste streams at the source

Systems without separation of wastewaters at the source manage the total mixture of wastes, including black water and grey water, and can be designed as conventional centralized systems, or less common distributed systems. The main features of centralized systems include the collection of wastewater in or near houses, transport by sewers to a treatment plant, and discharges of treated effluents to the receiving waters by outfalls. In many developing countries, centralized systems do not work well, because the facilities do not operate properly, or in some places not at all, due to the lack of management, maintenance, funding and training. The main research issues concerning centralized treatment include: further development and refinement of biological methods for nutrient removal from wastewater and the recovery of nutrients; the use of membrane technology for wastewater treatment; the development of anaerobic methods for sludge digestion and treatment; the incineration of sewage sludge (biosolids); and control of new chemicals of concern, such as endocrine disruptors, pharmaceuticals (including antibiotics), and personal care and therapeutic products (Marsalek et al., 2007).

In small towns, or suburban and peri-urban areas, distributed (local) systems are used in wastewater management and employ simple and inexpensive technologies to treat wastewater:

- Wastewater infiltration: After sedimentation, the wastewater is infiltrated into the soil in a constructed filter plant. Good reductions of organic matter, nutrients and bacteria may be achieved. This technology should not be used for highly polluted wastewaters or where underlying groundwater is used as a source of drinking water (Metcalf and Eddy, 2003).
- Constructed wetlands: Various forms of constructed wetlands are commonly used to treat wastewater or as a polishing step after a conventional treatment, mainly for nutrient removal. The major limitation of this technology is the need for a large space (Kadlec and Knight, 1996).
- Treatment ponds represent one of the most common treatment technologies used in both developed and developing countries. The main advantages of ponds include simple construction and relatively low capital costs, removals of solids and organic pollutants and pathogens, sludge digestion within the system, low maintenance, simple process operation, and good potential for resource recovery (Shilton, 2005). Disadvantages include large land footprints, algae growth contributing to increased Biogeochemical (or Biological) Oxygen Demand (BOD) and suspended solids (SS) concentrations in the effluent, effects of climatic conditions and algal blooms on performance, and inconsistent nutrient removal (Shilton, 2005).
- Anaerobic treatment technologies like anaerobic filters and especially upflow anaerobic sludge blanket (UASB) reactors have gained much popularity during the

BOX 4.2 Case study of Ho Chi Minh City, Viet Nam

The Binh Hung Canal is 4 km long and situated in the northeast part of Ho Chi Minh City, Vietnam. It serves a drainage area of 785 hectares (ha) and currently collects domestic wastewater from a population of about 120,000 people living in a poor area of the city. Furthermore, industrial and food processing operations, including textile-dying factories, seafood processing plants and paper mills, discharge unknown amounts of wastewater into the canal. This overload of pollutants results in fully anaerobic conditions in the canal and a bad odour. Because of the black appearance of the water, the canal is now called the Den (black) Canal.

Because of a lack of space caused by housing encroachment (photo below), the wastewater treatment plant has been designed as an aerated lagoon with stabilization ponds. The current design treatment capacity is 30,000 m³/day; the ponds occupy a surface area of about 22 ha.

Local residents benefit from this wastewater treatment plant not only because of the elimination of the odour and sanitation problems, but also due to the fact that the surroundings of the maturation ponds have been arranged as a public park.

Source: Nguyen et al., 2006. (See also colour plate 2)

Photo: Aerial view of the Den Canal wastewater treatment plant in Ho Chi Minh City (Vietnam). Picture courtesy of Jan van Lint (photo credit: Belgian Technical Cooperation).

past decade. They have two distinct advantages: the possibility of energy recovery through biogas production and a small footprint. The latter makes them particularly suitable for use in urban areas. UASB drawbacks include efficiency at high temperatures only, effluent requiring further treatment, and an explosion hazard caused by the captured and stored biogas.

4.4.3 Water and wastewater reuse for environmental benefits

The reuse of treated wastewater has been addressed as a practice reducing demands on potable water supply and water imports to urban areas. In this section, wastewater reuse is addressed as a water quality improvement measure reducing the discharge of pollutants into receiving waters. Such is the case, for example, in the State of Florida, where the Anti-degradation Policy for Reuse Projects prohibits new or expanded discharges for domestic wastewater treatment facilities unless the proponent can demonstrate that such discharges would be 'clearly in the public interest'. Moreover, the reuse of untreated wastewater is also practised in some regions of the world, due to the lack of water and economic resources to treat wastewater before reuse. Jiménez and Asano (2004) estimate that worldwide, at least 21 million hectares (ha) are irrigated with treated, diluted, partly-treated or untreated wastewater. In some arid or humid tropic countries, wastewater is used in urban agriculture for irrigation, mainly because of wastewater availability, local demand for fresh produce and the need to support people living on the verge of poverty. Smit and Nasr (1992) estimate that at least one-tenth of the world's population consumes crops irrigated with wastewater. Direct reuse of treated stormwater and wastewater has been applied in a number of European countries, mostly for garden irrigation and the flushing of toilets. Various types of waters, including reclaimed wastewater, are used in aquaculture to breed fish and grow aquatic crops. Some guidance for wastewater reuse can be obtained from the WHO guidelines (WHO, 1989), which are under review (Kamizoulis, 2005).

Recognizing high pollution loads carried by wastewater, and that even treated effluents cause impacts on receiving water quality, wastewater reuse as a control of environmental flows is an important measure applied in developed countries, and even where applied for other reasons (in regions with water shortages, or for agricultural irrigation in developing countries, for example), it produces the same benefits.

4.4.4 Technology and site selection

Alternative options for managing urban wastewater include both projects that aim to change the behaviour of citizens, as well as projects that focus on the construction of certain infrastructure or technologies (UNEP, 2003). Even if the infrastructure option has been chosen, it is still important to verify whether sufficient attention has been paid to the prevention of wastewater generation. If the answer is affirmative, the infrastructure for a certain treatment technology should be constructed.

The scale of the planned infrastructure is an important factor in the selection of a particular technology; some technologies are better suited to large scales, whereas others are better suited to small scales. Considerations for the selection of the project scale are as follows:

- Small scale systems inside the city reduce the costs for large infrastructural works for drainage and/or centralized wastewater treatment.
- Small scale systems allow the involvement of citizen groups in the design phase and in some cases in the operation of the systems.
- Small scale systems in the city improve the quality of the urban environment, by increasing green areas and by providing habitats for flora and fauna within the city, although the quality of such habitats need to be examined.

- Large scale systems benefit from economies-of-scale.
- Large scale systems allow for more efficiently organized operation and maintenance.

Once the scale has been defined, proper sites need to be identified on the basis of the following factors: availability and price of land, topography, soil type and capacity for infiltration or retention, groundwater table level, distance to residential areas (avoid odour nuisance), current land-use, possible multifunctional use (e.g., playground, recreation or a wildlife corridor), and discharge or reuse possibilities.

Once the site and scale have been chosen, the optimal technology must still be selected. The choice of technology also feeds back into the site selection; some technologies are only possible under certain site conditions. The choice among the various available technologies is best carried out using a set of previously defined criteria. The selection process should be as transparent and rational as possible. A common practice, however, is for utilities and consulting companies to select only 'proven technologies' for which they already have experience, or even off-the-shelf designs. This practice stifles innovation in urban water management.

Lists of factors that need to be taken into account when selecting a technology are available in the engineering literature (Metcalf and Eddy, 2003). One should keep in mind that for a comprehensive assessment of technologies, it is not sufficient to address only technical criteria; others, especially institutional capacity for operation and maintenance and financial sustainability, need to be taken into account as well. A comprehensive assessment would include health, environmental, social, economic and technical criteria (Veenstra and Alaerts, 1997; UNEP, 2003).

Quantification, or scoring of the selection criteria, needs to be done on the basis of the collected data, or when data are not available, on the basis of model simulations. After the criteria have been scored, a Multi Criteria Analysis (MCA) or another optimization method can be used to select the best option. MCA can be a simple procedure of giving weight to the criteria and calculating the total score for the options. The main purpose of the MCA is to make the decision process more transparent, so that different stakeholders can see how decision makers actually arrive at their decisions. The decision making process can also be supported by a number of software packages that are specially designed to facilitate technology selection. During a project's construction and after its completion, continual evaluations should be conducted to manage the project properly and ensure that the planning objectives will be met.

4.5 CONCLUDING OBSERVATIONS

Ecosensitive management of urban aquatic habitats represents a great challenge facing all environmental planners and urban water resources managers. The goal of this effort can be defined as preserving, rehabilitating or restoring aquatic habitats in urban waters, including streams, wetlands, rivers, lakes and estuaries. To achieve this goal, it is recommended to apply the concept of sustainable development, which can be implemented through the ecosystem approach based on four pillars – society, environment, economy and technical realisation – further supported by an appropriate institutional framework. In practical terms, such implementation is then based on multiple technical measures designed to counter the adverse effects of urbanization on aquatic habitat. These measures are considered in the provision of urban water services, which cause modifications of, and impacts on, aquatic habitats. In the field of water supply, the essential measures include managing the import of water into urban areas – excessive withdrawals affect flow regime in aquatic habitats and contribute to returns of large volumes of wastewater. The use of secondary sources of water (e.g., rainwater) and water and wastewater reclamation and reuse all offer good solutions conserving high quality water for appropriate uses. Urban drainage and flood protection exert pronounced impacts on aquatic habitats, including changes in flow regime, physical habitat structure, energy sources, chemical fluxes, and biotic interactions. The appropriate ecosensitive technical measures include advanced stormwater management, which employs measures for maintaining water balance at individual sites, thus reducing runoff generation and the mobilization of pollutants; reuse of rainwater and stormwater to the maximum extent; the use of larger structural measures, including ponds and wetlands to provide further controls and create new habitats; and measures in near-shore areas of receiving waters to mitigate stormwater effects on receiving waters. In the field of flood protection, there is a change in design philosophy by moving away, wherever feasible, from large structural measures (dams, reservoirs, dykes, etc.) to a greater use of non-structural measures (non-occupancy of flood plains; flood forecasting and warning, and evacuations; and flood insurance), or even to living with floods, whereby flood plains are used for land conservancy, wildlife habitat, and agricultural activities, including cattle grazing. Finally, the effects of wastewater management on aquatic habitats can be mitigated by reducing wastewater generation (e.g., by wastewater reclamation and reuse) and providing high levels of treatment. Some of these treatment technologies are applied as distributed local systems employing such systems as wastewater infiltration facilities, constructed wetlands, and treatment ponds. As emphasized in the section on ecosensitive strategies, holistic integrated approaches, encompassing all the technical measures listed here, need to be applied to achieve the overall goal of protecting aquatic habitats in urban waters.

REFERENCES

Alluvium, 2006. *Stormwater harvesting from waterways and drainage systems*. Report prepared for Melbourne Water, Australia.

Azzout, Y., Barraud, S., Cres F.N. and Alfakih, E. 1994. *Techniques alternatives en assainissement pluvial* (in French: Alternative techniques for stormwater drainage). Lavosier Tec and Doc, Paris, ISBN: 2-85206-998-9.

Baptista, M.B., de Oliveira Nascimento, N. and Barraud, S. 2005. *Tecnicas compensatorias em drenagem urbana* (in Portuguese: Compensation techniques in urban drainage). ABRH Press, Porto Alegre, Brazil.

Bergue J-M. and Ruperd Y. 1994. *Guide technique des bassins de retenue d'eaux pluviales*. Tec et Doc, Lavoisier, Paris.

Bishop, C.A., Struger, J., Barton, D.R., Shirose, L.J., Dunn, L, Lang, A.L. and Shepherd, D. 2000a. Contamination and wildlife communities in stormwater detention ponds in Guelph and the Greater Toronto Area, Ontario, 1997 and 1998. Part I – Wildlife communities. *Wat. Qual. Res. J. Canada* 35(3), 399–435.

Bishop, C.A., Struger, J., Shirose, L.J., Dunn, L. and Campbell, G.D. 2000b. Contamination and wildlife communities in stormwater detention ponds in Guelph and the Greater Toronto Area, Ontario, 1997 and 1998. Part II – Contamination and biological effects of contamination. *Wat. Qual. Res. J. Canada* 35(3), 437–74.

Bixio, D., De heyder, B., Cikurel, H., Muston, M., Miska, V., Joksimovic, D., Schafer, A.I., Ravazzini, A., Aharoni, A., Savic, D. and Thoeye, C. 2005. Municipal wastewater reclamation: where do we stand? An overview of treatment technology and management practice. *Wat. Supply* 5(1), 77–85.

Brown, R., Mouritz, M. and Taylor, A. 2005. Institutional capacity. T.H.F. Wong. (ed.) *Australian Runoff Quality: a guide to water sensitive urban design*, Melbourne, Australia.

CDM (Camp Dresser MacKee), Larry Walker Assoc., Uribe and Assoc., and Resources Planning Assoc. 1993. *California stormwater Best Management Practice handbook.* California Stormwater Quality Association, Menlo park, CA, US.

CIRIA. 2000a. *Sustainable urban drainage systems: a design manual for Scotland and N Ireland.* Report C521. Construction Industry Research and Information Association, London.

CIRIA. 2000b. *Sustainable urban drainage systems: a design manual for England and Wales.* Report C522. Construction Industry Research and Information Association, London.

CIRIA. 2001. *Sustainable urban drainage systems: best practice manual.* Report C523. Construction Industry Research and Information Association, London.

City of Cape Town. 2003. *Floodplain Management Guidelines.* Version 1. Produced by the Catchment, stormwater and river management branch of the City of Cape Town. Cape Town, South Africa.

Colas, H., Pleau, M., Lamarre, J., Pelletier, G. and Lavalle, P. 2004. Practical perspective on real-time control. *Wat. Qual. Res. J. Canada,* 39(4), 466–78.

Cowden, J.R., Mihelcic, J.R. and Watkins, D.W. 2006. Domestic Rainwater Harvesting Assessment to Improve Water Supply and Health in Africa's Urban Slums. Proc. of World Environmental and Water Resources Congress, May, 2006, Omaha, NE.

Day, E., Ractliffe, G. and Wood, J. 2005. *An audit of the ecological implications of remediation, management and conservation or urban aquatic habitats in Cape Town, South Africa, with reference to their social and ecological contexts.* Freshwater Consulting Group, Cape Town.

Day, J.A. 1987. Conservation and management of wetlands in the greater Cape Town area. R.D. Walmesley and M.L. Botten (eds). *Proc. of a Symp. on Ecology and Conservation of Wetlands in South Africa.* Occasional Report Series No 28, Ecosystem Programmes, Foundation for Research Development, CSIR, Pretoria, pp. 192–97.

DayWater Project 2003. *Review of the use of stormwater BMPs in Europe.* Report 5.1. http://daywater.enpc.fr/www.daywater.org/

Dunne, T. 1986. Urban Hydrology in the Tropics: problems solutions, data collection and analysis. *Urban climatology and its application with special regards to tropical areas,* Proc. of the Mexico Tech Conf., Nov. 1984, World Climate Programme, WMO, Geneva.

Engineers Australia 2005. *Australian Runoff Quality: a guide to water sensitive urban design.* Wong, T.H.F. (ed.), Melbourne, Australia.

Directive 2000/60/EC of the European Parliament and of the Council of 23 October 2000 establishing a framework for Community action in the field of water policy.

Exall, K. 2002. A Review of Water Reuse and Recycling, with Reference to Canadian Practice and Potential: 2. Applications. *Water Qual. Res. J. Canada,* 39(1), 13–28.

Friedler, E. and Hadari, M. 2006. Economic feasibility of on-site greywater reuse in multi-storey buildings. *Desalination,* 190, 221–34.

Geiger, W.F., Marsalek, J., Rawls, W.J. and Zuidema, F.C. (eds). 1987. *Manual on drainage in urban areas. Volume I – Planning and design of drainage systems.* Studies and reports in hydrology No. 43, UNESCO, Paris, ISBN 92-3-102416-7.

Ghisi, E. and Ferreira, D.F. 2007. Potential for potable water savings by using rainwater and greywater in a multi-storey residential building in southern Brazil. *Building and Environment*.

Gijzen, H.J. 1997. Duckweed based wastewater treatment for rational resource recovery. II Symposia Internacional sobre Ingenieria de Bioprocesos, Mazatlan, Mexico, 8–12 September 1997, pp. 39–40.

Goddard, M., 2006. Urban greywater reuse at the D'LUX Development. *Desalination*, 188, 135–40.

Horner, R.R., Skupien, J. J., Livingston, E.H. and Shaver, H.E. 1994. *Fundamentals of urban runoff management: technical and institutional issues*. Terrene Institute, Washington DC.

Horner, R., May, C., Livingston, E., Blaha, D., Scoggins, M., Tims, J. and Maxted, J. 2002. Structural and nonstructural BMPs for protecting streams. B.R. Urbonas (ed.) *Linking Stormwater BMP Designs and Performance to Receiving Water Impact Mitigation*, American Society of Civil Engineers, New York, NY, pp. 60–77.

Jeppesen, B. 1996. Domestic greywater re-use: Australia's challenge for the future. *Desalination*, 106, 311–15.

Jiménez, B., Chávez, A., Maya, C. and Jardines, L. 2001. Removal of microorganisms in different stages of wastewater treatment for Mexico City. *Wat. Sci. Tech.*, 43(10), 155–62.

Jiménez, B. and Asano, T. 2004. Acknowledge all approaches: The Global outlook on reuse. *Water 21*, December, pp. 32–37.

Jones, R.C., Via-Norton, A. and Morgan, D.R. 1997. Bioassessment of BMP effectiveness in mitigating stormwater impacts on aquatic biota. L.A. Roesner (ed.) *Effects of Watershed Development and Management on Aquatic Ecosystems*, American Society of Civil Engineers, New York, NY, pp. 402–17.

Kadlec, R.H. and Knight, R.L. 1996. *Treatment Wetlands*. Lewis Publishers, Boca Raton, FL, pp. 893.

Kamizoulis, G. 2005. The new draft WHO guidelines for water reuse in agriculture. *Technical workshop: The Integration of reclaimed water in water resource management*. Lloret de Mar, Costa Brava, Girona, Oct. 209–20.

Leopold, L. 1968. *Hydrology for urban land planning – a guidebook on the hydrologic effects of urban land use*. U.S. Geol. Survey Circular 554, USGS, Reston, VA.

Lloyd S.D. 2004. *Exploring the opportunities and barriers to sustainable stormwater management practices in residential catchments*. PhD Thesis, Department of Civil Engineering, Monash University, Melbourne, Australia.

Lloyd S., Francey M. and Skinner L. 2004. Cost of incorporating WSUD into Greenfield site development and application of an offset trading scheme. *Proceedings for the WSUD Conference*, Adelaide, South Australia, Australia.

Lundqvist, J., Narain, S. and Turton, A. 2001. Social, institutional and regulatory issues. C. Maksimovic, and J.A. Tejada-Guibert (eds) *Frontier in urban water management: Deadlock or hope*. IWA Publishing, London, UK.

MacRae, C.R. 1997. Experience from morphological research on Canadian streams: Is control of the two-year frequency runoff event the best basis for stream channel protection? L.A. Roesner (ed.) *Effects of Watershed Development and Management on Aquatic Ecosystems, Proceedings of an Engineering Conference*, ASCE, pp. 144–62.

Marsalek, J., Barnwell, T.O., Geiger, W. F., Grottker, M., Huber, W.C., Saul, A.J., Schilling, W. and Torno, H.C. 1993. Urban drainage systems: design and operation. *Wat. Sci. Tech.* 27(12), 31–70.

Marsalek, J. and Chocat, B. 2002. International report: Stormwater management. *Wat. Sci. Tech.*, 46(6–7), 1–17.

Marsalek, J., Jimenez-Cisneros, B.E., Karamouz, M., Malmquist, P.-A., Goldenfum, J. and Chocat, B. 2007. *Urban Water Cycle Processes and Interactions*. Taylor & Francis Group, London, UK.

Marsalek, J., Watt, W.E. and Anderson, B.C. 2006. Trace metal levels in sediments deposited in urban stormwater management facilities. *Wat. Sci. Tech.*, 53(2), 175–83.

Melbourne Water 2005. *WSUD Engineering Procedures: Stormwater.* CSIRO Publishing, Melbourne, Australia.

Metcalf and Eddy Inc. 2003. *Wastewater engineering: treatment, disposal and reuse.* G. Tchobananoglous and F.L. Burton (eds), McGraw-Hill series in water resources and environmental engineering, New York, US.

Ministry of Environment (MOE) 2003. *Stormwater management planning and design manual.* Ontario Ministry of the Environment, Toronto, Ontario.

New South Wales Health Department 2005. *Greywater and sewage recycling in multi-unit dwellings and commercial premises – Interim guidance.* Circular No. 2005/051, New South Wales Health Department, Gladesville, NSW, Australia (http://www.health.nsw.gov.au/policies/GL/2005/pdf/GL2005_051.pdf), last accessed on Feb. 19, 2007.

Nguyen, T.P.T., Chu, Q.H., Dieu, H.D., Rousseau, D.P.L., Goethals, P.L.M., De Pauw, N., Trinh, T.T.T. and Vasel, J.-L. 2006. The WWTP of Tan Hoa – Lo Gom canal in Ho Chi Minh City, Vietnam. *Proceedings 7th IWA International Conference on Waste Stabilization Ponds,* 25–27 September 2006, Bangkok, Thailand.

Nhapi, I. and Gijzen, H.J. 2005. A 3-step strategic approach to sustainable wastewater management. *Water SA,* 31(1), 133–40.

Nolde, E. 1999. Greywater reuse systems for toilet flushing in multi-storey buildings – over ten years experience in Berlin. *Urban Water,* 1, 275–84.

Parkinson, J. and Mark, O. 2005. *Urban stormwater management in developing countries.* ISBN: 1843390574, IWA Publishing, London, UK.

Research Wise 2006. *Summary of water quality community perceptions.* A synopsis of qualitative research on behalf of Melbourne Water, Melbourne, Australia.

Riordan, E.J., Grigg, N.S. and Hiller, R.L. 1978. Measuring the effects of urbanization on the hydrologic regimen. P.R. Helliwell (ed.), *Urban Drainage, Proc. Int. Conf. on Urban Storm Drainage, Southampton,* April 1978, pp. 496–11.

Roesner, L.A., Bledsoe, B.P. and Rohrer, C.A. 2005. Physical effects of wet weather discharges on aquatic habitats – present knowledge and research needs. E. Erikson H. Genc-Fuhrman, J. Vollertsen, A. Ledin, T. Hvitved-Jacobsen and P.S. Mikkelsen (eds), *Proc. 10th Int. Conf. on Urban Drainage* (CD-ROM), Copenhagen, Denmark, Aug. 21–26, 2005.

RossRakesh, S., Francey, M. and Chesterfield, C. 2006. Melbourne Water's stormwater quality offsets. *Australian Journal of Water Resources,* 10, 241–50.

Ruiter, W. 1990. Watershed: Flood protection and drainage in Asian Cities. *Land and Water International,* 68, 17–19.

Schilling, W. 1989. *Real time control of urban drainage systems – the state of the art.* IAWPRC Scientific Report No.2, Pergamon Press, ISBN: 0-08-040145-7, London.

Schueler, T.R. 1987. *Controlling Urban Runoff: A Practical Manual for Planning and Designing Urban BMPs.* Washington Metropolitan Water Resources Planning Board, Washington DC.

Schueler, T.R. 1992. Mitigating the adverse impacts of urbanization on streams: a comprehensive strategy for local government. P. Kimble and T. Schueler (eds) *Watershed Restoration Sourcebook.* Publication # 92701 of the Metropolitan Washington Council of Governments.

Shilton, A. (ed.) 2005. *Pond treatment technology.* ISBN: 1843390205, IWA Publishing, London, UK.

Smit, J. and Nasr, J. 1992. Urban agriculture for sustainable cities: using wastes and idle land wand water bodies as resources. *Environment and Urbanization,* 4(2), 141–52.

Sriyaraj, K. and Shutes, R.B.E. 2001. An assessment of the impact of motorway runoff on a pond, wetland and stream. *Environment International,* 26, 433–39.

Sweeney, R.A. 1995. Rejuvenation of Lake Erie. *Earth and Environmental Science,* 35(1), 65–66.

Tangsubkul, N., Moore, S. and Waite, T.D. 2005. Incorporating phosphorus management considerations into wastewater management practice. *Environmental Science and Policy,* 8, 1–15.

Tarczynska, M., Frankiewicz, P. and Zalewski, M. 2002. The regulation and control of hydrologic and biotic processes within reservoirs for water quality improvement. UNEP - IETC. *Guidelines for the Integrated Management of the Watershed – Phytotechnology and Ecohydrology.* IETC Technical Publication Series.

Tucci, C. and Villanueva, A. 2004. Land use and urban floods in developing countries. A. Szollosi-Nagy and C. Zevenbergen, C. (eds), *Urban flood management.* ISBN 04 1535 998 8, Taylor and Francis Publishers, UK.

UNEP. 2000. *International source book on environmentally sound technologies for wastewater and stormwater management.* UNEP, Osaka, Japan (available at http://www.unep.or.jp/ietc/Publications/TechPublications/TechPub-15/main_index.asp) visited on Feb. 17, 2007.

UNEP. 2003. *Managing urban sewage: An introductory guide for decision makers.* UNEP, Osaka, Japan (available at: http://www.unep.or.jp/ietc/Publications/Freshwater/FMS10/index.asp) visited on Feb. 17, 2007.

UNESCO-IHE, UNEP/GPA. *Wastewater technologies and management for Pacific Islands.* Training course on CDROM.

Urbonas, B. and Jones, J.E. 2002. Summary of emergent urban stormwater themes. B.R. Urbonas (ed.), *Linking Stormwater BMP Designs and Performance to Receiving Water Impact Mitigation,* American Society of Civil Engineers, New York, NY, pp. 1–8.

Urbonas, B. 1994. Assessment of Stormwater BMPs and Their Technology. *Wat. Sci. Tech.,* 29(1–2), 347–53.

US EPA. 2000. *Low Impact Development (LID).* A Literature Review. Office of Water (4203) EPA-841-B-00-005. Environmental Protection Agency Washington DC 20460 October 2000.

Van Betuw, W. 2005. *Evaluated source separating wastewater management systems.* MSc thesis Wageningen University.

Veenstra, S. and Alaerts, G.J., 1997. *Technology selection for pollution control.* IHE International Institute for Infrastructural, Hydraulic and Environmental Engineering. Delft, The Netherlands.

Victorian Stormwater Committee, 1999. *Urban stormwater best practice environmental management guidelines.* CSIRO Publishing, Melbourne.

Walsh, C.J., Leonard, A.W., Ladson, A.R. and Fletcher, T.D. 2004. *Urban stormwater and the ecology of streams.* Cooperative Research Centre for Freshwater Ecology and Cooperative Research Centre for Catchment Hydrology, Canberra.

Water Environment Federation (WEF) and American Society of Civil Engineers (ASCE) 1998. *Urban runoff quality management.* WEF Manual of Practice No. 23, ASCE Manual and Report on Engineering Practice No. 87, WEF, Alexandria, VA, USA.

Water Pollution Control Federation (WPCF) 1989. *Combined sewer overflow pollution abatement.* Manual of Practice FD-17, WPCF, Alexandria, VA.

WHO, 1988. *Urbanization and its implications for child health: Potential for action.* World Health Organization. Geneva, Switzerland.

WHO, 1989. *Health guidelines for the use of wastewater in agriculture and aquaculture.* Report of a WHO Scientific Group, WHO Technical Report Series 778. WHO, Geneva, Switzerland.

WHO and UNICEF, 2000. *Global water supply and sanitation assessment 2000 report.* WHO, Geneva, Switzerland.

Winblad, U. and Simpson-Hebert, M. (eds), 1998. *Ecological sanitation.* Stockholm Environment Institute, Stockholm, Sweden.

Yoder, C. 1989. The development and use of biocriteria for Ohio surface waters. Flock, G.H. (ed.), *Proc. National Conf. Water Quality Standards for the 21stCentury,* US EPA, Office of Water, Washington DC, pp. 139–46.

Chapter 5

Aquatic habitat rehabilitation: Goals, constraints and techniques

Kinga KRAUZE[1], Marek ZAWILSKI[2] and Iwona WAGNER[1]

[1] European Regional Centre for Ecohydrology under the auspices of UNESCO, Polish Academy of Sciences, 3 Tylna Str, 90–364 Lodz, Poland
[2] Technical University of Lodz, Department of Environmental Engineering, 6 Al. Politechniki, 90–924 Lodz, Poland

Integration of aquatic habitats with the city infrastructure without considering their ecological properties results in their serious degradation. Therefore, according to the Water Framework Directive (WFD) of the European Union, (Directive 2000/60/EC), urban aquatic habitat's are often classified as heavily modified or artificial water bodies. Although their restoration is often not possible (see Chapter 1), preserving or restoring habitats physical features and thus ecological functions of ecosystems nonetheless increases the likelihood of their sustainable functioning. Habitat rehabilitation strengthens the defence mechanisms of ecosystems and enhances their resilience to human impact. It allows for maintaining ecological quality within the variety of functions that urban aquatic habitats have to play in an urban environment. Moreover, as further explained in Chapters 6 and 7, rehabilitated habitats incorp-orated into the city landscape provide services to society more efficiently.

This chapter focuses on the practical aspects of urban aquatic habitat rehabilitation, with a special focus on rivers, and provides generalities and techniques for management of their hydrological dynamics and biotic structure.

5.1 ASSESSING THE ECOLOGICAL POTENTIAL OF THE RIVER

5.1.1 Buffering mechanisms

Urbanization deprives rivers of their natural external protection. Landscape modifications and the use of resources affect the river catchment and thus water resources. Proper land-use planning and mitigation policy applied in the early stages of the urbanization process may, however, favour maintenance of some regulatory potential of ecosystems.

The ecological potential of a river is dependent on the presence of numerous buffering (or defence) mechanisms. These emerge from the structure of the river channel itself (interplay between biota and environment), as well as from its valleys and floodplains. The stabilization of ecosystem functions occurs through the regulation of such processes as nutrient fluxes, erosion and sediment transport, light access and hence the rate of primary production, biomass removal from the system and its external deposition, capacity for sediment and chemicals immobilization, biodiversity and efficiency of energy flow. The defence mechanisms depend to a great extent on the habitat structure, and its ability to support aquatic life and maintain the dynamics of natural processes in the ecosystem. The better the defence mechanisms are developed, the higher the resilience of the ecosystem, hence its capacity for disturbance assimilation (see also Chapters 2 and 6). Therefore, a proper assessment of an aquatic ecosystem quality increases chances for selecting a suitable management and rehabilitation strategy.

5.1.2 Methods for river state assessment

5.1.2.1 Bioassessment

For decades the focal point of river assessment has been the analysis of chemical and physical water parameters. Well-known, easily applicable and precise methods allowed continuous monitoring of a number of variables, which were considered as critically important in the case of urban rivers. Norms specifying the tolerable contamination levels allowed for the improvement of water security and setting sensitive early-warning systems.

The new strategy of environmental management shifts attention from maintaining good water quality towards maintaining the ecosystem value as a whole and revealed the limitation of the earlier techniques. Water quality is not necessarily a sign of good ecological status of a river. It is especially obvious in the cases where the use of modern, efficient treatment and well implemented environmental laws led to significant improvements of the chemical water quality, while the habitat structure and biological diversity remained very poor. Furthermore, when rehabilitation is a management target, the urgent demand for more comprehensive monitoring and evaluation of the ecosystem state is an undisputable fact. The evaluation method needs to be based on a component of the aquatic system that responds to overall conditions of a whole river ecosystem. That is why new assessment techniques focus on the structure of biocenosis, which reflects the quality of water and aquatic habitat as well as the complex biotic interactions between species. This approach is currently supported by the EU's WFD and the Habitat Directive (Council Directive 92/43/EEC) of the European Commission and international regulations, such as the Convention on Biological Diversity (CBD) and the Ramsar Convention.

The conceptual background for modern river monitoring programmes was provided by the theory of ecological integrity. It states that an ecosystem maintains its integrity only when the pattern of internal and external processes and interactions between ecosystem attributes produce the biotic community corresponding to the natural state of the region-specific habitats (Karr, 1981). That resulted in the development of biological monitoring and bioassays, based on phytoplankton, phytobenthos, macrophytes, benthic invertebrates and fish as indicators of the environmental status.

The selection of an indicator group of organisms brings about differences in accuracy and errors that might occur in the evaluation process as a consequence of different life strategies of organisms. Indicator variability decreases in the following direction: phytoplankton > zooplankton > macroinvertebrates > macrophytes > fish. This means that an assessment based on phytoplankton reaction, which exhibits high seasonal variability, will be more frequently erroneous and have a lower statistical power than the one using fish as an indicator (see Figure 5.1.; Lapińska, 2004). The best results, however, can be obtained by using various indicator groups simultaneously. This follows from the fact

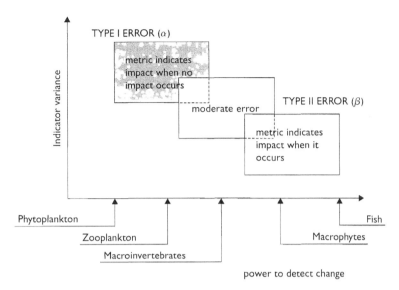

Figure 5.1 Bioassessment: A conceptual model presenting errors assessment for different indicator groups

Sources: Lapińska, 2004; Johnson 2001.
A type I error (α) is made in testing an hypothesis when it is concluded that a result is positive when it really is not, while a type II error (β), when it is concluded that something is negative when it really is positive.

that some groups are effective as early indicators, while others are effective as late-warning indicators, and their effectiveness depends on the type of stressor. Thus the selection of complementary early- and late-warning indicator groups increases the probability of detecting an impact if it occurs, and hence has become a common practice. For example, macrophytes, which have low seasonal variability but exhibit slow changes in the community structure, are useless as an early warning indicator. But, for the same reason, when change is detected in the macrophyte species composition, then the probability of no impact occurrence is low (Johnson, 2001). A highly variable phytoplankton community and the periphyton group are excellent indicators of nutrient enrichment, as they tend to respond very rapidly to changes in the water trophic state. Response from macro invertebrates is not that rapid, but they are more sensitive to habitat characteristics (also more habitat-bound) and to long-term trophic changes. Statistically, the most accurate fish indicators are not useful in the case of nutrient enrichment incidents, but they are the most appropriate if the ecosystem stressor is temperature or chemical contaminant.

There are three main methodological approaches used for riverine quality bioassessment:

● A single metric approach based on a single parameter from an indicator group; for example, species richness, density of individuals, similarity or diversity of communities (Saprobic index, Trent Biotic Index, the Danish Stream Fauna Index, the Belgian Biotic Index, the Extended Biotic Index)

- A multi-metric approach, which aggregates several metrics; for example, the Index of Biotic Integrity for macro-invertebrates or for fish
- A multivariate approach based on measures of the mathematical relationships among samples (similarity in structure of two communities) for two or more variables (for example, qualitative presence/absence of species, or quantitative abundance or biomass of species) – Jaccard's similarity coefficient, cluster analysis, discriminant analysis, ordination techniques, generalised linear models, logistic regression, and Bayesian models (Lapińska, 2004; Dahl, 2004). The choice between approaches and their application depends on the anticipated number of stressors affecting the system and the questions which need to be answered.

5.1.2.2 Physical and geomorphological assessment

As presented earlier in Chapter 2, anthropogenic pressures on habitats are classified into five major categories according to the physical and geomorphologic features being affected. All the impact categories, including flow regime, habitat structure, water quality, food sources and biotic interactions should be considered prior to preparing a river management and rehabilitation plan. As habitat quality provides a template for biological processes and ecosystem dynamics, many bio-assessment methods have already incorporated physical assessment protocols in order to describe habitat conditions of indicator biota groups, e.g. river assessment systems like the United Kingdom's System for Evaluating Rivers for Conservation (SERCON) or The River Habitat Survey (RHS).

These physical assessment methods include the evaluation of the geo-morphological characteristics of the river bed and valley, the distribution of habitats within the river channel (riffles, runs and pools), the presence and variety of patches of uniform substrate, vegetation and flow velocity and light access, and finally the preservation of longitudinal river characteristics (zonation). It can also help in deciding between management options (as presented in Chapter 1) of degraded ecosystems (see Table 5.1).

5.2 TECHNIQUES IN URBAN RIVER REHABILITATION

Ecosensitive measures for the environmental management of urban areas support river restoration remediation and rehabilitation, wherever conditions are favourable. They also recognize the role of river corridors in maintaining and enhancing biodiversity and ecosystem services and improving human well-being. Ecohydrology being a complementary approach proposes as a long-term rehabilitation target a gradual increase in the assimi-lative capacity of urban ecosystems as a result of integrated activities at the scales of the catchment, valley and aquatic habitat. It includes re-establishing structures and processes stabilizing ecosystem functions – the continuity of flows between the river and its surroundings, the continuum of river habitats when and where feasible, the rehabilitation of wetlands and buffer strips – through modification of the hydrological regime. In this process, 'green feedbacks' – regulatory properties of plant cover (Zalewski et al., 2003), are utilized for stabilizing the microclimate, water conditions, soil properties and enhancing plant succession.

Because of heavy modifications of urban river catchments, it is unrealistic to restore the cycling of water and nutrients to a state resembling the natural one. However, catchment

Table 5.1 Methods of physical and geomorphological assessment

Method	Characteristics	Link to biota
Geomorphic River Styles	– based on the theory of geomorphological processes – enables to predict future channel character and its response to disturbances	– between geomorphology and biota with respect to habitat
State of the River Survey	– assessment at several levels: catchment, river sections, tributaries, using data components individually or together	– between the parameters measured and stream biota (substrate, riparian vegetation)
River Habitat Survey (RHS)	– assessment of river habitat quality based on its physical structure – uses a database with habitat requirements, site classification and association of flora and fauna with different habitats – 500 metre long sites are randomly selected with 50 m intervals in between; 10 spot checks are performed and numerous features are recorded – can be linked with RIVPACS and SERCON.	– on the basis of the biotope and functional habitat approach – Biotope approach – use of habitat units by biota is inferred from the known physical conditions – functional habitat approach – the habitat is defined from the knowledge of inhabiting biota
The Integrated Habitat Assessment System (IHAS)	– measures components of stream habitats relevant to macroinvertebrates: substrate, vegetation, physical conditions – assessment is based on rating and scoring components in order to derive the continuum of habitat quality	– assumption of habitat units relevant to the macroinvertebrate occurrence
The Instream Flow Incremental Methodology (IFIM)	– computer models and analytical procedures designed to predict changes in fish habitat due to flow alterations – software includes: Physical Habitat Simulation System, Legal Institutional Analysis Model, and Physical Habitat Assessment Model, Stream Network Temperature Model and System Impact Assessment Model	– assuming that flow-dependent habitat and water temperature determine the carrying capacity of rivers for fish

Sources: Newson et al., 1998; Dunn, 2000; Phillips et al., 2001; Parson et al., 2002.

structure development has to be considered in rehabilitation plans for urban water bodies and wastewater systems (Geiger and Dreiseitl, 1995). Any credible historical information on the former course of the river and its natural reservoirs should be analysed as an indispensable condition for successful rehabilitation and restitution of the basic river attribute – connectivity with its valley and the catchment. The medium integrating all elements of the system, and also the technical elements with green spaces, is water.

5.2.1 Rehabilitation of hydrological dynamics of river habitats

As mentioned in Chapter 1, not every urban river can be restored. The best results are achieved by the rehabilitation of moderately impacted urban rivers that flow through green or residential areas. But many urban streams, converted into urban channels and drains receiving stormwater, combined sewer overflow (CSO) discharges and treated sewage, cannot change their present function any more. The examples of such situations can be found in Chapter 9 in the case studies of Lyon, France and Lodz, Poland. Moreover, the conversion of such 'channels' into open watercourses is usually impossible due to their location in densely built-up city districts. A proper master plan should therefore select for rehabilitation river segments that still have some ecological and aesthetic values, favourable location and ecological potential to undergo the rehabilitation process. Making a distinction between functional channels and potentially restorable rivers also allows for additional management of stormwater peak flows by diverting extreme flows from the restored river and directing them into a separate channel system, thereby protecting aquatic habitats against pollution and hydraulic overloading. On the other hand, rehabilitated urban rivers can be also used for the following:

- transporting relatively clean stormwater
- relieving some combined and storm sewers and treatment plants from wet-weather peak flows and hydraulic loads, so that the less polluted rainwater is not discharged into the urban sewers and allowed to overload municipal treatment plants
- reducing the risk of flooding in surrounding areas, because of more attenuated flow in rehabilitated riverbeds (Stecker, 1996; Zawilski et al., 1998; Zawilski, 2001).

Ideally, restored aquatic habitats should be connected to separate storm sewers, designed to drain catchments producing limited surface runoff pollution. This helps avoid expensive investments in protection against pollution. The use of roof runoff, draining into a river through a system of natural or even artificially created small watercourses with landscaped ponds and wetlands, can be one idea (Conradin and Buchli, 2004). Runoff from roofs does not require any pre-treatment, unless the roofing material produces pollutants, which may be the case for roofs made from metal sheets containing zinc or copper. In other cases, especially where runoff from streets and parking lots is used, the separation of mineral particles and oil derivatives from runoff is obligatory, except in low density residential areas with single family units. Recently, green roofs have also been introduced, with the purpose of retaining most rainwater completely, and although this option cannot be widely applied in cities, it may improve runoff dynamics and quality considerably.

The following sections list aspects that should be taken into consideration, from a technical point of view, by master plans for the regulation of hydrological dynamics and the physical structure of aquatic habitats.

5.2.1.1 Attenuation of peaks flows using in-catchment or on-watercourse storage reservoirs

In an optimal solution, one would restore the flow regime as close as possible to the pre-urbanization natural level. This would require either attaining a very low runoff coefficient for the catchment or a high throttling of the catchment outflow, which is equivalent to providing considerable runoff storage. Storing stormwater in the catchment is a better solution and can be achieved with the use of numerous local detention elements

(Andoh et al., 2001). Storing stormwater in the catchment itself is preferable for water quality management. Since rainwater usually has a relatively good quality, it gets polluted only during the passage along the city surface by collecting contaminants along the way. Increasing water retention and reducing surface runoff achieves therefore not only reduced stormwater flows and flood risks, protecting the habitats and associated ecosystems from physical damage, but also improves the quality of the receiving waters.

The required specific storage volume should be about 40 to 100 cubic metres per hectare (m³/ha) of impervious surface, or even more where possible. This volume guarantees a radical decrease of peak stormwater flow (Zawilski and Sakson, 2002; 2004). Sometimes, the desired attenuation of peak flows may be achieved in detention ponds situated on streams (Zawilski, 2001). Usually the sufficient water surface area of such ponds equals a few percent of the connected catchment area. However, in the case of extreme storms, storage of the total runoff in such ponds is not possible, and emergency overflows or bypasses have to be activated (see Figure 5.2).

High flows of a ten-year return period may be diverted to a parallel channel of a sufficient cross-section or to a large floodplain area.

In general, there is also the possibility of diverting stormwater into the ground by using constructed infiltration facilities, which help reduce the overall effective stormwater runoff (Geiger and Dreiseitl, 1995). Such a method may be applied in favourable geological conditions (i.e., with a minimum 1 m layer of sandy soils above the groundwater table and no risk of water seepage into basements). If possible, every rainwater roof leader should be diverted onto grass rather that being directed into a storm sewer. It should be recognized that unfavourable soil conditions (soils with low percolation rates) do not prevent the use of underground storage of stormwater completely, because one can use optional storage in underground units filled with gravel and allowing slow exfiltration of stormwater, as for instance done in the so-called Mulden-Rigolen-systems used in Germany (Geiger and Dreiseitl, 1995). The use of underground storage and

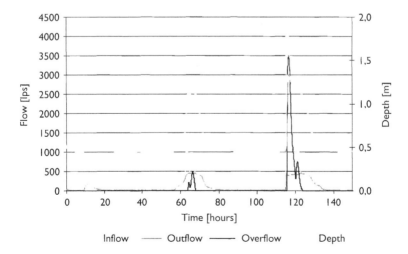

Figure 5.2 Balancing urban stream flow with an in-line pond of 6,000 m² bottom area and 20,940 m³ maximum active volume

infiltration can cause a distinct reduction of stormwater peak flows – by as much as 50% compared to traditional solutions.

A similar effect can be obtained by using alternative land covers – from completely impervious to porous, or those with larger openings. Good results are always achieved by increasing the extent of green areas (parks, land-water buffer zones, vegetation patches, etc.). This measure increases water interception on plant surfaces, evapotranspiration, improves the structure of soils by increasing the organic matter content and thus enhances water retentiveness and reduces runoff.

5.2.1.2 Managing the high flow regime in floodplains and riverbeds

Floodplains often disappear during urban development and the associated transformation of riverbeds. For restored urban rivers, the common flows can be balanced, but the river corridor still may be seen as a flood channel, whose flow capacity should not be exceeded. Therefore, floodplain rehabilitation as a river corridor collecting and conveying water during heavy rainfalls is beneficial. Such floodplains, however, are difficult to recover, due to existing buildings, a lack of space and technical problems. In land-use planning, they should be kept as open land without any development and be protected against possible damage.

It should be underlined that the hydraulic capacity of a restored river channel is often smaller than that of a regulated one, because the friction factor (roughness) of natural cover of the riverbed (gravel, stones, brush and grass) is higher than that of concrete. Consequently, design calculations should be performed to establish design values of water depth, flow velocity and general capacity of the restored stream. This also points to the need to address the whole system, the catchment and the stream, to reduce outflows from the catchment and demands on the restored stream capacity.

In the case of larger and deeper riverbeds transporting high flows, a reinforcement of the riverbed may be unavoidable in order to avoid erosion damage of the bed and banks. Combinations of natural and artificial materials as well as plant species adapted to local hydrological conditions (flow velocity, hydro-period, etc.) help maintain ecosystem functions. Ripraps with in-situ soil-on-gravel filter layers and geotextile, micro-piles, willow bundles and shrubs placed above the maximum water level can also be used (Pagliara and Chiavaccini, 2004; Urbonas et al., 2004). Similar special reinforcements should be placed around sewer outlets.

5.2.1.3 Assurance of minimum flows during dry weather

In the case of some small urban streams, there is a danger of flow disappearance during longer periods of dry weather, particularly if situated in a highly impermeable catchment, disconnected from floodplains, close to a stream spring and/or after larger storm and combined sewer outlets have been disconnected. Such a change in a habitat's major feature destabilizes the ecological processes in rivers and reduces their functioning and ability to provide services, such as self-purification. Storing stormwater along the river or on the catchment surface can partially soften the problem and maintain river flow for a period of days or even weeks. Also, the river course may be supplied by the interflow of groundwater originating from storm sewers. Connecting drainage pipes collecting groundwater, like foundation drains, or any other type of clean water (from water treatment plants or industry, for example) should be considered. In extreme cases, supplying rivers by pumped groundwater is possible, but should not be a common solution.

A system's ability to retain water will largely depend on geomorphology and soil properties. A restored riverbed with sandy soils and low groundwater tables will not be tight enough to keep a permanent dry weather flow. One can consider making the riverbed more watertight by using a geo-membrane or natural loam/clay materials; however, if there is a possibility of water exchange between the local groundwater and the riverbed, the application of artificial bed seals should not be used, or at least only considered with caution.

5.2.1.4 Flattening of the river longitudinal grade

Sills placed in rivers with higher gradients can decrease their longitudinal gradient, maximum velocity and erosion. They can preserve the physical structure of habitats during high flows, as well as during long-lasting droughts, by increasing water retention in the area. Special attention should be paid to the proper construction of sills, eliminating the possibility of bottom scouring (Adduce et al., 2004). In order to preserve the natural (unmodified) habitat structure, interference of anthropogenic elements, such as dams, culverts, sills, weirs, and pipes should be minimized, and as much as possible, grade controls should mimic natural features and be made of natural materials. (Geiger and Dreiseitl, 1995; Stahre, 2002; Urbonas et al., 2004).

5.2.2 Rehabilitation of the physical structure of river habitats

There are two key practices applied in channel reconstruction: maintaining hydraulic connections (partially discussed in the previous section) and stream meanders. Maintenance of hydraulic connectivity allows the movement of water and biota between the stream, abandoned channel arms and adjacent floodplain areas. It prevents losses of aquatic habitat areas and their diversity. Backwater areas adjoining the main channel can potentially be used for spawning and rearing for many fish species and are a key habitat component for wildlife species that live in, or migrate through, the riparian corridor. Stream meander rehabilitation is considered when transforming a straightened stream channel in order to reintroduce natural dynamics, improve channel stability, habitat quality, aesthetics, and other stream corridor functions or values. This enables the creation of a more stable stream with more habitat diversity, but also requires an adequate area, and consequently, adjacent land-use may constrain this practice in some locations. Constituting an integral part of hydrological management, the rehabilitation of the habitat's physical structure will be given special attention here, because of its particular importance for supporting aquatic life. Its diversity depends on the creation of variable conditions reflecting those found in natural microhabitats supporting a wide variety of organisms that should be also included in the rehabilitation masterplan.

5.2.2.1 Re-meandering straight watercourse sections

Straight channels are typical for regulated urban riverbeds, and for the sake of hydrological regime rehabilitation and support of the aquatic life, their extent should be minimized by retaining them only where necessary. Re-meandering a riverbed helps to restore the flow regime to semi-natural characteristics and enables the attenuation of

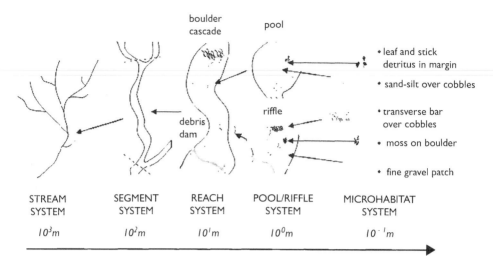

boulder
cascade

pool

• leaf and stick
  detritus in margin

• sand-silt over cobbles

riffle

• transverse bar
  over cobbles

debris
dam

• moss on boulder

• fine gravel patch

STREAM SYSTEM	SEGMENT SYSTEM	REACH SYSTEM	POOL/RIFFLE SYSTEM	MICROHABITAT SYSTEM
$10^3 m$	$10^2 m$	$10^1 m$	$10^0 m$	$10^{-1} m$

Figure 5.3 Hierarchical organization of a stream system and its subsystem habitats (See also colour plate 3)

Source: modified from Frissell et al., 1986.

the flow by reducing water velocity. Retrofit construction of pool-riffle structures (see Figure 5.3) favours biodiversity by providing variable conditions in the river cross-section (Dale, 1996; Schwartz et al., 2002), and thus stabilizes ecosystem functioning and increases resistance to, and resilience in dealing with, anthropogenic impacts.

In naturally meandering watercourses, riffles (shallow zones) and pools (deep zones) form a regular pattern. However, habitats in natural streams can be even more complex at this level, including other habitat forms like rapids, runs, falls, and side channels. The lowest level in this hierarchy is a microhabitat system of relatively homogenous substrate type, water depth and velocity (see Figure 5.4). If possible, channel morphology should be reconstructed with respect to the presented catchment context, with appropriate meso- and micro–features, in order to shape hydraulic properties and provide a template for physical (sedimentation, erosion), chemical (accumulation, sorption) and biological (self-purification, production, denitrification) processes typical for a particular system. Without a proper template, the recovery of the ecosystem, with its complexity and resilience, is not possible (Zalewski and Naiman, 1985).

To create riffles, runs, flats, glides and open pools, all of which are important components of the physical structure of river meso-habitat rehabilitation, one can use several practices including the following:

–   **Boulder Clusters:** placed in the baseflow channel, where they provide cover, create scour holes, or refuges with reduced velocity
–   **Weirs or Sills:** log, boulder, or quarry stone structures placed across the channel and anchored to the stream bank and/or bed to create pool habitat, hydraulic diversity in uniform channels and control bed erosion and deposition
–   **Fish Passages:** in various forms enhance the opportunity for target fish species to freely move to upstream areas for spawning and other life functions

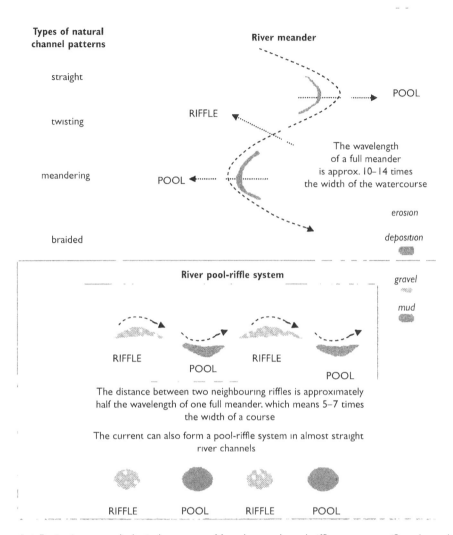

Figure 5.4 Basic river morphological structure: Meanders and pool-riffle sequences (See also colour plate 4)

Source: Lapińska, 2004: modified from Calow and Petts, 1992.

- **Log/Brush/Rock Shelters:** structures installed in the lower portion of stream banks to enhance fish habitat, encourage food web dynamics development, prevent stream bank erosion, and provide shading
- **Lunker Structures:** usually cells constructed of heavy wooden planks and blocks that are embedded into the toe of stream banks at the channel bed level to provide covered compartments for fish shelter and habitat and prevent stream bank erosion
- **Migration Barriers:** prevent undesirable species from accessing upstream areas and thereby effectively serve specific fishery management needs

- **Tree Cover:** may be created by placing felled trees along the stream bank to provide overhead cover, aquatic organism substrate and habitat, and deflect stream current and thereby prevent scouring, or sediment deposition, and drift
- **Wing Deflectors:** structures protruding from either stream bank but not extending fully across the channel, which deflect flows away from the bank and scour pools by constricting the channel and accelerating flow
- **Grade Control Measures:** typically implemented as rock, wood, earth, and other material structures placed across a channel and anchored to the stream banks to provide a 'hard point' in the streambed that resists the erosion forces in the zone of degradation, and/or serves to reduce the upstream energy slope to prevent bed scour. These structures are used to stop cutting in degrading channels in order to improve bank stability, increase low water depths for the upstream habitat, and serve as a possible low-flow migration barrier (FISRWG 10, 1998).

5.2.2.2 Stream bank management and maintenance

Stream bank treatment consists of numerous techniques, which serve to protect river banks and simultaneously support some river defence strategies. It includes the establishment of bank vegetation, which serves as a filter for chemicals, a soil stabilizer, biomass interception, grazing and regulator of predator-prey interactions, etc. (see Table 5.2).

The restored riverbed requires maintenance in order to keep its flow conveyance capacity, e.g., by removing wood debris and sediment, controlling invasive plant species and conducting renovation works. From a legal point of view, each water reservoir should have its banks kept free of buildings, fences and other anthropogenic elements that impair free access.

5.2.3 Reconstruction of biotic structure

Considerations of the role of vegetation in the functioning of riverine systems and the rebuilding of biotic structures should include 3 steps:

- establishing macrophyte communities in river beds
- structuring plant cover on river banks
- developing plant vegetation in river valleys.

In urban catchments, the re-establishment of biotic structure is usually limited to rehabilitating banks and in-channel vegetation, which will be described in more detail in the following two subsections. Renewing valley vegetation is often more difficult, due to space and land-use constraints. Still, it should be recognized that valley vegetation is important for water dynamics and quality, playing a role in dissipating wave energy during flood events, stabilizing river discharge and mitigating the effects of floods and droughts. It also provides a number of transitional land-water habitats and supports the development of biodiversity in adjacent areas. By providing a framework for biogeochemical processes, it also takes part in biogeochemical processes in land-water interface zones and enhances matter retention and self-purification in rivers. A partial recovery of valley functions can be based on the development or use of existing recreational/green areas; other functions should be harmonized with engineering systems including sewerage, water

Table 5.2 Techniques for restoring the physical structure of river banks

	Description	Application
Live Stakes	Live, woody cuttings tamped into the soil to root that grow and create a living root mat stabilizing the soil by reinforcing and binding soil particles together, and by extracting excess soil moisture.	Effective where site conditions are uncomplicated, construction time is limited, and an inexpensive method is needed. Appropriate for repair of small earth slips and slumps that are frequently wet. Can be used to stake down surface erosion control materials. Requires toe protection where toe scour is anticipated.
Live Fascines	Dormant branch cuttings bound together into long sausage-like, cylindrical bundles and placed in shallow trenches on slopes to reduce erosion and shallow sliding.	Can trap and hold soil on stream bank by creating small dam-like structures and reducing the slope length into a series of shorter slopes. Facilitate drainage when installed at an angle on a slope. Enhance conditions for colonisation of native vegetation.
Log, Root Wad, and Boulder Revetment	Boulders and logs with root masses attached, placed in and on stream banks to provide stream bank erosion protection, trap sediment, and improve habitat diversity.	Will tolerate high boundary shear stress if logs and root wads are well anchored. Suited for streams where fish habitat deficiencies exist.
Rip-rap	A blanket of appropriately sized stones extending from the toe of slope to a height needed for long term durability.	Appropriate where long term durability is needed, design discharges are high, there is a significant threat to life or high value property, or there is no practical way to otherwise incorporate vegetation into the design. Can be vegetated (see joint plantings). Commonly used to form of bank protection.
Stone Toe Protection	A ridge of quarried rock or stream cobble placed at the toe of a stream bank as armour to deflect flow from the bank, stabilize the slope and promote sediment deposition.	Should be used on streams where banks are being undermined by toe scour, and where vegetation cannot be used. Stone prevents removal of the failed stream bank material that collects at the toe, allows re-vegetation and stabilizes stream banks.
Tree Revetments	A row of interconnected trees attached to the toe of a stream bank or to dead heads in a stream bank to reduce flow velocities along eroding stream banks, trap sediment, and provide a substrate for plant establishment and erosion control.	Works best on streams with stream bank heights under 3.6 m and bank-full velocities under 1.8 m per second. Captures sediment and enhances conditions for colonization by native species particularly on streams with high bed material loads.
Vegetated Geogrids	Alternating layers of live branch cuttings and compacted soil with natural or synthetic geotextile materials wrapped around each soil lift to rebuild and vegetate eroded stream banks.	Quickly establishes riparian vegetation if properly designed and installed. Can be installed on a steeper and higher slope and has a higher initial tolerance of flow velocity than brush layering.

Source: Lapińska, 2004; FISRWG 10, 1998.

supply, etc., as discussed in Chapter 4. Whenever natural communities may be still present in the urban landscape, it is recommended, for ecological, economic and aesthetic reasons, to preserve and enhance them. While this is usually rare, there are often remnants of the natural plant cover in a form of scarce patches. They are important indicators of the continuing connection between river and valley, and hence should be considered when planning rehabilitation projects.

5.2.3.1 In-stream vegetation: The use of aquatic plants

The role of aquatic plants is the most pronounced in small- and medium-size streams or rivers, in which water depth does not exceed 2 m during the highest discharges. In urban rivers, aquatic plants play a role of channel hydraulics and flow moderators and provide refuges for biota. They also regulate sedimentation and nutrient retention. On the other hand, one has to consider that the use of aquatic vegetation in management can be restricted by habitat conditions, including high peak discharges, high water turbidity (hydraulic tension, low light), and also by demands imposed on the river as a component of the water management system. Aquatic plants may impede flow velocity, which leads to rising water levels, and, due to decomposition and matter interception, they may reduce the channel cross-sectional area needed for flow transport. Thus, a proper introduction of vegetation for sustainable river management requires the following:

- precise calculations of flow characteristics;
- a good understanding of channel hydraulics;
- knowledge about biomass distribution, and
- a good understanding of ecology of the dominant plant species (Krauze, 2004).

Plant growth and expansion rates are regulated by temperature, light access, flow distribution in the channel, nutrient and oxygen concentrations. A careful planning of river channel morphology helps to define the distribution of aquatic plants through analyses of river hydraulics. Temperature and light access can be regulated through composing riparian structure and building up a canopy of shrubs and trees wherever macrophyte growth is to be limited. However, it should be considered that some plant species may pose health hazards, e.g., *Heracleum mantegazzianum* (giant hogweed) and *Conium maculatum* (hemlock). Others can be difficult to control, e.g., *Reynoutria japonica* (Japanese knotweed), *Stratiotes aloides* (water soldier), *Impatiens glandulifera* (Himalayan balam), and *Nymphoides peltata* (fringed water lily). *Phragmites australis* (Norfolk reed) and *Typha latifolia* (bullrush), which require more space, are suitable only for larger rivers (NRA Severn-Trent Region), and should not be used for habitat improvement in smaller streams.

Establishing aquatic plants in urban streams requires the application of special procedures and techniques in order to prevent plant damage and wash out in the early stages of succession. Such procedures include the creation of low flows, partially isolated zones within the river bed, and preparation of the substratum for planting (see Figure 5.5).

In cases where some intervention in aquatic plant zones is necessary in order to protect the functions of river ecosystems, it is important to consider the sequential exploitation of such zones, making sure that, wherever possible, sections of the river, or at least parts

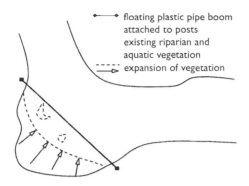

floating plastic pipe boom
attached to posts
existing riparian and
aquatic vegetation
expansion of vegetation

Notch planting of rooted plants, rhizomes and marginal plants, and introduction of
floating leaved or submerged plants

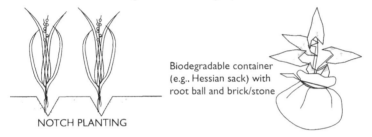

NOTCH PLANTING

Biodegradable container
(e.g., Hessian sack) with
root ball and brick/stone

Figure 5.5 Enhancement of riparian and aquatic vegetation by protection from waves and current action

Source: Krauze, 2004; modified from Cowx and Welcomme, 1998 and NRA Severn- Trent Region.

of the middle and edge sections, are left undisturbed and can act as refuges for plants and animals and allow re-colonization. The timing of plant harvesting, transplantation for retaining a river's profile and protecting its fish and invertebrates during all maintenance operations should be maintained in order to assure a fast system recovery.

5.2.3.2 Bank and riparian vegetation. The role of land/water ecotones

Riparian areas are complex systems characterized by floral and faunal communities distinctly different from surrounding upland areas. There are many different types of riparian ecotones: swamp forests, vegetated banks, meadows, littoral zones, marshes, floating mats, oxbow lakes, etc. Their common feature is occasional flooding. The water regime modifies the rates of aerobic and anaerobic biochemical processes and hence seasonal releases and the removal of phosphorus and nitrogen. Structuring plant cover on river banks further ensures bank protection against erosion, increases channel roughness and regulates matter transport. It also increases infiltration of the surface runoff and provides a framework for biogeochemical processes taking part in land-water interface zones, thereby enhancing stream self-purification.

The crucial role of land/water ecotones in the protection of river systems and regulation of in-stream processes is underlined by Naiman et al. (1989). They report that riparian areas are major determinants of water and nutrient flows across the riverine landscape. According to Petersen et al. (1992), the efficiency of wide ecotone zones (19–50 m) may reach even 78 to 98% removal for N in surface waters and 68 to 100% in groundwater. Other authors estimate reduction efficiencies at a level of 50 to 90% for nitrogen and 25 to 98% for phosphorus (in groundwater), depending on initial concentrations, buffer zone width, soil type, ecotone slope and interactions between plants and other organisms. The most intense pollution reduction occurs within the first 10 m of an ecotone zone. The influence of ecotone vegetation on water quality is possible only when the connection between terrestrial and water ecosystems is maintained. The basis for this connection is water circulation and is significant only along non-regulated river courses and reservoir shores.

Using natural vegetation for trapping sediments, nutrients and toxicants can threaten healthy functioning and have detrimental effects on wildlife and people (see Chapter 8). Thus, in conditions of no natural buffering zones or of high pollution loads, it may be necessary to create artificial riparian zones, which can remove much of the chemical and sediment loads from runoff before such loads reach the main water body or an area of special ecological or social interest. The factors that have to be considered before preparation of an action plan are as follows:

- the geomorphology of the area
- hydrological dynamics, e.g., water level fluctuations, the timing and range of extreme events
- plant species composition in natural land/water ecotones in the area
- species-specific efficiency of nutrient removal, growth rate and decomposition
- interactions between plant species
- planned use of the area (for recreation, agriculture, etc.).

5.2.4 Phytoremediation

Considering the problems with predicting metabolic pathways and transforming chemical compounds carried with wastewaters and stormwater in runoff, in some cases the above measures may be enhanced by implementing phytoremediation measures in catchments and river valleys. Phytoremediation refers to a variety of cost-effective methods of soil, groundwater, surface water and air remediation using plants. It usually concerns the upper 50-cm deep layer of soil when herbaceous plants are used (Kucharski et al., 1998; Raskin and Ensley, 2000), or deeper when deep-rooting trees are used to extract organic solvents from deeper aquifers (Negri et al., 1996).

There are several methods of phytoremediation classified according to the biochemical processes involved, application method and the type of plant used:

- phytoextraction or phytoconcentration: the contaminant is concentrated in the roots, stem and foliage of the plant
- phytodegradation: plant enzymes help catalyse the breakdown of the contaminant molecule
- rhizosphere biodegradation: plant roots release nutrients to micro-organisms which are active in biodegradation of the contaminant molecule
- volatilization: transpiration of organics, selenium and mercury run through leaves of the plant

- stabilization: the plant converts the contaminant into a form which is not bioavailable, or the plant prevents the spreading of a contaminant plume (UNEP, 2003).

Phytoremediation appears to be a 'natural technology', i.e., relatively simple. There are, however, some important factors that should be observed carefully in order to achieve the expected results and avoid disappointments:

- plant species used for phytoremediation should be selected appropriately
- indigenous species locally adapted and resistant to the substances polluting the soil should be given preference
- optimally, the plant should not require special care but should be tolerant to natural variability of weather conditions and capable of adapting to the characteristics of the remediated aquatic habitat (e.g., flow and hydroperiod). ·

5.2.5 Increasing capacity of urban habitats for water and nutrients' retentiveness

Wetlands' capacity for water and pollutants retention depends on two components. The first one is a hydrologic assimilative capacity, related to the retention and infiltration of surface water inputs. This is why extreme hydrological conditions, stormwater inputs and droughts, have to be considered in the process of wetland design. They have to be sufficiently large to retain certain volumes of water at depths and durations adjusted to the hydroperiod tolerated by vegetation, which may vary considerable for some species (Hammer, 1992; Taylor, 1992). On the other hand, the role of oxygen concentration, which depends to a great extent on hydroperiod, in trapping efficiency for different compounds must also be considered. Almendinger (1999) demonstrates that the most permanent removal of phosphorus occurs in deeper wetlands and ponds where phosphorus is scavenged by algae and accumulated in an organic form in sediments. Moreover, wetlands should be permanently inundated to maintain anaerobic conditions and hence minimize the decay of organic matter and support denitrification.

The second component of assimilative capacity is the chemical assimilative capacity, which consists of macrophyte uptake, microbial transformation and chemical sorption by bed sediments. It is determined by hydrological regime, sedimentation rate and soil processes, but also depends on the dynamics of biota. According to Devito and Dillon (1993) assimilation is low when chemical input exceeds metabolic rates of organisms, thus it is inversely correlated to runoff and coincides with high biotic assimilation rates during the growing season. Consequently, significant differences in wetland efficiency may be observed (Vought et al., 1995; Kadlec and Knight, 1996).

At the catchment scale, the assimilative capacity of a river system is a function of the number and area of biogeochemical barriers efficient in nutrient storage. The function of biogeochemical barriers in cities should be provided by green areas and rehabilitated river valleys, which have to be integrated with the city's sewerage system. The specific components include sedimentation ponds, rainwater collectors, stormwater bypasses, drainage ditches for surface flow collection, tree and shrub zones, grassland areas, planted woodlands, floating and submerged macrophytes zones, embankments, etc. The sequence of such elements depends on local demands. As a general rule, however, vegetation is considered as a key element of biogeochemical barriers. In order to achieve their high efficiency, five major properties of the planned system have to be specified, as well as their spatial-temporal variability:

- the type of plant community to be created and native species present in the area
- mechanical and physical characteristics of individual species
- substrate properties and species requirements
- changes of river discharges and maximum water depth during flooding
- seasonality of biological processes
- types of stress imposed on the system (toxicity of chemicals, concentration of salts, extreme flows, seasonal drying, high sediment load, use by people, etc.).

Although the benefits of using constructed wetlands to reduce the chemical load in municipal wastewaters have been well documented (Mitsch and Jorgensen, 1989; 2004; UNEP-IETC), it is important to underline that their construction has to be preceded by developing a detailed understanding of the pattern of biological, physical and chemical processes occurring in the planned wetland. As the wetland ability to reduce river pollution by anthropogenic contaminants results from complex redox reactions and microbial processes, eventually, chemical transformations may lead to more toxic and bioavailable forms of some chemicals (Shiaris, 1985). It is especially true for the areas exposed to mixtures of chemicals, which often occurs in industrial cities. Forstner and Wittman (1981) show that anoxic conditions lead to the reduction of arsenic and chromium to more toxic states. Helfield and Diamond (1997) provide evidence that alternating oxygen conditions, due to periodic inundation and drying of wetland, may enhance the bioavailability of metals adsorbed to hydrous oxides of iron and manganese. Oxygen conditions may also influence the effects of metals on biota through the enhancement of methylation. In this case, high levels of microbial activity in wetlands result in net methylation and subsequent biomagnification of mercury (Helfield and Diamond, 1997; Portier and Palmer, 1989; Wood et al., 1968).

Figure 5.6 presents an example of the connection of different elements of a rehabilitated river valley, including rehabilitation of physical structure of aquatic habitats, reestablishment of vegetation cover and its integration with a sewer system.

5.3 IMPROVING THE LIKELIHOOD OF SUCCESS IN THE IMPLEMENTATION OF REHABILITATION PROJECTS

In order to achieve long-term success, the rehabilitation of urban river systems has to address both the symptoms and causes of ecological disturbances. The source of disturbances is often removed in time and space from the target system, as urban rivers constitute an integral part of complex wastewater, stormwater and combined sewage systems established and developed in the past.

There are four main stages in a rehabilitation project (see Figure 5.7):

- establishing a vision
- developing a plan
- implementing the plan
- monitoring and conducting a project review.

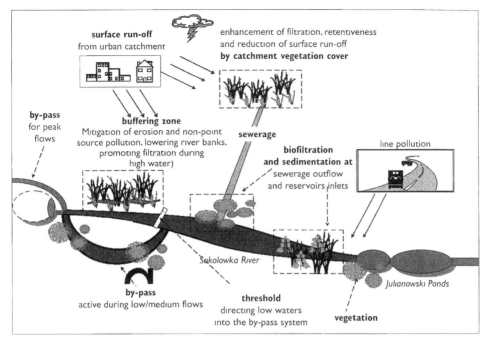

Figure 5.6 Proposed elements of a rehabilitated urban river valley and aquatic habitat (See also colour plate 5)

Source: Bocian and Zawilski, 2005.

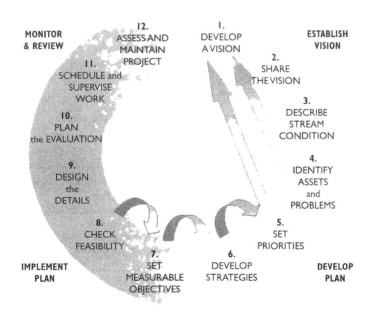

Figure 5.7 A 12-Step rehabilitation procedure

Source: Modified from Rutherfurd et al., 1999.

Project development should involve a learning-by-doing procedure of adaptive management. To avoid failure, a rehabilitation project should start with specifying the goals and objectives, explicit enough to become a basis for a project evaluation checklist used towards the project's end and in the post-rehabilitation period. It should also include a consensus of all the involved parties on the rehabilitation programme goals, a capacity building plan for the required expertise and scope, and analysis of uncertainties related to the project realization, among others.

The key issue in the pre-rehabilitation period is choosing a proper planning scale. The project should cover an area large enough to reduce boundary effects and address at least some sources of disturbances of the system, and what controls them, in an efficient way. The area should also enable proper monitoring of rehabilitation results and the achievement of intermediate objectives, which is crucial during the rehabilitation project. In the case of failure, the necessary actions should be planned to correct the problems. The evaluation should also include verification of performance indicators, whether they appear to be insufficient or inappropriate considering the project goals.

Finally, during the post-rehabilitation period, it is inevitable to compare the achievements against the planned objectives. One of the major purposes of this exercise is to assess the similarity of the restored system to the target one, as well as its sustainability. Other criteria include the assessment of all critical components of the restored system, clarification of ecological, economic and social benefits achieved by rehabilitation, and the project's cost-effectiveness. An important issue is establishing the project schedule, which makes it possible to check rehabilitation results against some unusual environmental conditions, such as floods and droughts.

Generally speaking, there are several main causes of failure for rehabilitation projects. The first relates to a lack of institutional agreements and consensus among the stakeholders on rehabilitation goals. This can jeopardize the execution of the action plan or impose constraints on some necessary actions. The latter concerns improper implementation of the project, often resulting from a lack of a sufficient database, or using inadequate techniques, and/or a lack of mid-term assessment and improvement procedures. The third common reason of failure is an improper formulation of objectives and goals,

Table 5.3 Examples of conflicting goals in urban river rehabilitation projects

Socio-economic goals, awareness, & perception	Ecological targets
– increased risk of flooding after river re-meandering	– rehabilitation of river bed geometry and hydrological patterns
– unlimited access to water	– protection of river banks and riparian vegetation
– water for domestic and industrial use	– limited water uptake
– river damming for flood reduction and increased water storage	– dam elimination for rehabilitation of hydrological patterns
– development of water-sports and recreational areas	– creation of habitats for wildlife
– increased risk of water-related diseases	– re-establishing natural buffering structures – backwaters, wetlands, riparian strips
– high aesthetic value expected	– rehabilitation of 'original' or similar vegetation structure
– introduction of attractive species	– rehabilitation of the natural species structure

meaning that the rehabilitation project does not meet ecological or social expectations. Therefore discrepancies between goals, expectations and risk perception related to the involved actors (see Table 5.3) have to be considered and reconciled during the planning, implementing and monitoring of the rehabilitation project. Several of these conflicting goals can be addressed at the level of the urban area development planning.

REFERENCES

Adduce, C., Larocca, M. and Sciortino G. 2004. Local scour downstream of grade control structures in urban stream restoration. Enhancing Urban Environment by Environmental Upgrading and Rehabilitation. J. Marsalek (ed.) *NATO Science Series, IV. Earth and Environmental Sciences*, vol. 43, pp. 307–18.

Almendinger, J.E. 1999. A method to prioritize and monitor wetland restoration for water-quality improvement. *Wetlands Ecology and Management* 6: 241–51.

Andoh, R., Faram, M., Sephenson, A. and Kane, A. 2001. A novel intergrated system for stormwater management. Proc. of the NOVATECH conference, Lyon, vol.1, pp.391–98.

Bocian, J. and Zalwilski, M. 2005. Ecohydrology concept – merging the ecology and hydrology for successful urban stream rehabilitation. Proceedings of the Urban River Rehabilitation Conference, Dresden, September 2005.

Calow, P. and Petts, G.E. (eds). 1992. *The Rivers Handbook. Volume One. Hydrological and Ecological Principles*. Blackwell Science Ltd.

Conradin, F. and Buchli, R. 2004. The Zurich stream day-lighting program. Enhancing Urban Environment by Environmental Upgrading and Rehabilitation. J. Marsalek (ed.) *NATO Science Series, IV. Earth and Environmental Sciences*, vol. 43, pp. 277–88.

Council Directive 92/43/EEC of 21 May 1992 on the conservation of natural habitats and of wild fauna and flora.

Cowx, I.G. and Welcomme, R.L. 1998. Rehabilitation of rivers for fish. Fishing News Books. Oxford.

Dahl, J. 2004. Detection of Human-Induced Stress in Streams. Comparison of Bioassessment Approaches using Macroinvertebrates. PhD thesis. Department of Environmental Assessment, Uppsala.

Dale, A. 1996. Engineering implications of rehabilitation of urban channels. Proc. of the 7th Int. Conf. On Urban Drain., Hannover, vol.II, pp.1211–16.

Devito, K.J. and Dillon, P.J. 1993. The influence of hydrologic conditions and peat anoxia on the phosphorus and nitrogen dynamics of a conifer swamp. *Water Resources Research* 29: pp. 2675–85.

Directive 2000/60/EC of the European Parliament and of the Council of 23 October 2000 establishing a framework for Community action in the field of water policy.

Dunn, H. 2000. Identifying and protecting rivers of high ecological value LWRRDC Occasional Paper No.01/00.

FISRWG 10. 1998. Stream Corridor Rehabilitation: Principles, Processes, and Practices. The Federal Interagency Stream Rehabilitation Working Group (FISRWG) GPO Item 0120-A; SuDocs No. A 57.6/2:EN3/PT.653. http://www.usda.gov/stream_restoration.

Forstner, U. and Wittman, G.T.W. 1981. *Metal pollution in the aquatic environment*. Springer-Verlag, New York.

Frissell, C.A., Liss, W.J., Warren, C.E. and Hurley, M.D. 1986. A hierarchical approach to classifying stream habitat features: viewing streams in a watershed context. Environmental Management 10: pp. 199–214.

Geiger, W.F. and Dreiseitl, H. 1995. Neue Wege für das Regenwasser – Handbuch zum Rückhalt und zur Versickerung von Regenwasser in Baugebieten, Oldenbourg Verlag, München.

Hammer, D.A. 1992. Creating freshwater wetlands. Lewis Publishers, Chelsea, Michigan.

Helfield, J.M. and Diamond, M.L. 1997. Use of constructed wetlands for urban stream restoration: a critical analysis. *Environmental Management* 21(3): pp. 329–41

Johnson, R.K. 2001. Indicator metrics and detection of impact. K. Karttunen (ed.) *Monitoring and assessment of ecological status of aquatic environments*. Nordic Council of Ministers, TemaNord 563, pp: 41–44.

Kadlec, R.H. and R.L. Knight. 1996. Treatment Wetlands. Lewis Publishers, Boca Raton, FL, pp. 893.

Karr, J.R. 1981 Assessment of biotic integrity using fish communities. Fisheries Bethesda 6: pp. 21–27.

Krauze, K. 2004. Land-water interactions: How to Assess their Effectiveness. LAND-water interactions: Reduction of Contamination Transport. M. Zalewski and I. Wagner-Lotkowska (eds) *Integrated Watershed Management – Ecohydrology & Phytotechnology – Manual*. UNESCO IHP, UNEP-IETC, 75–97, pp. 169–88.

Kucharski R., Sas-Nowosielska A., Malkowski E. and Pogrzeba M. 1998. Report prepared for the U.S Department of Energy. Integrated approach to the remediation of heavy metal-contaminated land. IETU Katowice.

Lapińska, M. 2004. Streams and rivers: Defining their Quality and Absorbing Capacity. Management of streams and rivers: how to enhance absorbing capacity against human impacts. M. Zalewski and I. Wagner-Lotkowska (eds) *Integrated Watershed Management – Ecohydrology & Phytotechnology – Manual*. UNESCO IHP, UNEP-IETC, pp. 75–97, 169–88.

Mitsch, W.J. and Jørgensen, S.E. 1989. *Ecological Engineering: An Introduction to Ecotechnology*. John Wiley & Sons, New York.

Mitsch, W.J. and Jørgensen, S.E. 2004. *Ecological Engineering and Ecosystem Restoration*. John Wiley & Sons, New York.

Naiman, R.J., Decamps, H. and Fournier, F. (eds). 1989. *Role of land / inland water ecotones in landscape management and restoration, proposals for collaborative research*. UNESCO, Vendome, France.

Negri, M.C., Hinchman, R.R. and Gatliff, E.G. 1996. Phytoremediation: using green plants to clean up contaminated soil, groundwater, and wastewater. Proceedings, International Topical Meeting on Nuclear and Hazardous Waste Management, Spectrum 96. Seattle, WA, August 1996. American Nuclear Society.

Newson, M.D., Harper, D.M., Padmore, C.L., Kemp, J.L. and Vogel, B. 1998. A cost-effective approach for linking habitats, flow types and species requirements. *Aquatic Conservation: Marine and Freshwater Ecosystems* 8: 431–46.

NRA Severn – Trent Region F.R.C.N. Guidelines. Odum E. P. 1971. Fundamentals of Ecology. Saunders, Philadelphia.

Pagliara, S. and Chiavaccini, P. 2004. Urban stream restoration structures. Enhancing Urban Environment by Environmental Upgrading and Rehabilitation, J. Marsalek (ed.) *NATO Science Series, IV. Earth and Environmental Sciences*, vol. 43, pp. 239–52.

Parson, M., Thomas, M. and Norris, R. 2002. Australian River Assessment System: Review of Physical River Assessment Methods – A Biological Perspective. *Monitoring River Health Initiative Technical Report 21*, Environment Australia.

Petersen, R.C., Petersen, L.B.-M. and Lacoursiere, J. 1992. A building-block model for stream restoration. P.J. Boon, P. Calow and G.E. Petts (eds) *River Conservation and Management*. John Wiley & Sons, Chichester, New York, Toronto, Singapore, pp: 293–09.

Phillips, N., Bennett, J. and Moulton, D. 2001. Principles and tools for the protection of rivers, Queensland Environmental Protection Agency report for LWA.

Portier, R.J. and Palmer, S.J. 1989. Wetlands microbiology: Form, function, processes. D.A. Hammer (ed.) *Constructed wetlands for wastewater treatment: Municipal, industrial and agricultural*. Lewis Publishers, Chelsea, Michigan, pp. 89–105.

Raskin, I. and Ensley, B. (eds). 2000. *Phytoremediation of Toxic Metals*. Wiley Interscience NY.

Rutherfurd, I.D., Jerie, K. and Marsh, N. 1999. A rehabilitation manual for Australian streams, Vol. 1–2, Land and Water Resources Research and Development Corporation & CRC for Catchment Hydrology, Canberra.

Schwartz, J.S., Herricks, E.E., Rodriguez, J.F., Rhoads, B.L., Garcia, M.H. and Bombardelli, F.A. 2002.Physical habitat analysis and design of in-channel structures on a Chicago, IL urban drainage: a stream naturalization design process. Proc. 9th Int. Conf. on Urban Drainage, Portland, (9th ICUD) 2002. A CD-ROM collection, ASCE, US.

Shiaris, M.P. 1985. Public health implications of sewage applications on wetlands: microbiological aspects. P.J. Godfrey, E.R. Kaynor, S. Pelczanski and J. Benfordo (eds). *Ecological considerations in wetlands treatment of municipal wastewaters*. Van Nostrand Reinhold, New York, pp. 243–56.

Stahre, P. 2002. Integrated Planning of Sustainable Stormwater Management in the City of Malmo, Sweden. Proc. 9th Int. Conf. on Urban Drainage, Portland, (9th ICUD) 2002. A CD-ROM collection, ASCE, US.

Stecker, A. 1996. Steps to a new dewatering system in the Emscher-area. Proc. of the 7th Int. Conf. On Urban Drain., Hannover, vol.II, pp.1181–92.

Taylor, M.E. 1992. Constructed wetlands for stormwater management: A review. The Queen's Printer for Ontario, Toronto, Ontario.

UNEP. 2003. Phytotechnologies. A Technical Approach in Environmental Management. UNEP, Division of Technology, Industry and Economics. Freshwater Management Series No.7.

Urbonas, B.R., Kohlenberg, B., Thrush, C. and Hunter M. 2004. Restoring natural waterways in Denver, USA area. Enhancing Urban Environment by Environmental Upgrading and Rehabilitation, J.Marsalek (ed.) *NATO Science Series, IV. Earth and Environmental Sciences*, vol. 43, pp. 227–38.

Vought, L., Pinay, G., Fuglsang, A. and Ruffinoni, C. 1995. Structure and function of buffer strips from a water quality perspective in agricultural landscapes. *Landscape and urban planning* 31: pp. 323–31.

Wood, M.M., Kennedy, F.S. and Rosen, C.G., 1968. Synthesis of methyl-mercury compounds by extracts of a methanogenic bacterium. *Nature* 220:pp. 173–74.

Zalewski, M. and Naiman, R. 1985. The regulation of riverine fish communities by a continuum of abiotic – biotic factors. J.S. Alabaster (ed.) *Habitat modification and freshwater fisheries*. FAO UN Butterworths Scientific. London, pp. 3–9

Zalewski, M., Santiago-Fandino, V. and Neate, J. 2003. Energy, water, plant interactions: 'green feedback' as a mechanism for environmental management and control through the application of phytotechnology and ecohydrology. *Hydrological Processes* 17: 2753–67.

Zawilski, M. 2001. Management of urban stormwater and storm overflow water with the use of small natural watercourses: a case study of Lodz. Proc. of the NOVATECH' 01.

Zawilski, M., Kujawa, I., Zalewski, M. and Bis, B. 1998. Stormwater management and renovation of natural watercourses in Lódź. Proc. of the NOVATECH' 98.

Zawilski, M. and Sakson, G. 2002. Potential of alternative technologies concerning reduction of urban stormwater flow. Proc. 9th Int. Conf. on Urban Drainage, Portland, (9th ICUD) 2002. A CD-ROM collection, ASCE, USA.

Zawilski, M. and Sakson, G. 2004. Possibility and effect of OSR/OSD implementation on a densely built-up urban catchment. Proc. of the 6th Int. Conf. On Urban Drainage Modelling (UDM'04), Dresden, Sept. 15–17th, 2004, pp. 573–80.

Chapter 6

Ecohydrology of urban aquatic ecosystems for healthy cities

Maciej ZALEWSKI[1,2] and Iwona WAGNER[1,2]

[1] European Regional Centre for Ecohydrology under the auspices of UNESCO, Polish Academy of Sciences, 3 Tylna Str, 90-364 Lodz, Poland
[2] Department of Applied Ecology University of Lodz, 12/16 Banacha Str, 90-237 Lodz, Poland

As presented in Chapters 4 and 5, properly managed and rehabilitated urban aquatic habitats can be integrated into city infrastructure, operating as one system. A system can be defined as an organized integrated whole, made up of diverse, interrelated and interdependent parts. In the context of this definition, the twenty-first century urban environment is still very often far from being perfect, considering the integration of its anthropogenic functions with ecosystem functioning. Until recently, aquatic ecosystems and green areas in cities have often been considered as independent and relatively expensive sub-systems. This chapter suggests taking a broader look at the urban environment in the context of its integration with city functions. Urban ecosystems are considered here as vital assets, providing not only environmental values, but also services for society, thus improving the quality of life and development of the city of the future. This chapter also proposes an alternative perception of the interplay between the aquatic habitat (especially its hydrological properties) and biota, not only as a target of protection, but also as a tool for better water management and for enhancing habitats' defence mechanisms against urban impact.

6.1 INTRODUCTION

From the point of view of environmental science, urban environment can be considered as a highly condensed anthropogenic system, which is organized for the efficient flow of water, matter, energy and information. This extremely dynamic super organism can efficiently provide the services required by society, such as safe drinking water, efficient sanitation and mitigation of floods which is fundamental due to very high population density. However, increases in society's education and environmental awareness raises the public demand for improvement of the quality of life. Therefore other expectations, depending to a great extent on proper ecosystem functioning, arise. These include ecosystem services like those determining human health and quality of life, based on water quality improvement by self-purification and clean air, as well as those fulfilling materialistic and spiritual aspirations, such as high quality living spaces, recreational areas and aesthetic value. The services depend to a great extent on the functioning of aquatic ecosystems and their ability to cope with high impacts (see Chapter 2), determined, among others, by the size and distribution of green areas (see also Chapter 7). However, the limited availability and high price of land in cities make maximizing environmental amenities at low management a real challenge for any society. Therefore one alternative solution is increasing the absorbing capacity of ecosystems in order to improve their ability for coping with the highly condensed human impacts in urban areas. The methods for achieving this are implicitly provided by principles of ecohydrology (Zalewski et al., 1997), which suggest using 'ecosystem properties as a management tool' to enhance the efficiency of some regulatory processes. The solutions have to be integrated into the city 'system' by their harmonization with engineering solutions (Zalewski, 2000).

6.2 ECOHYDROLOGY CONCEPT AND PRINCIPLES

6.2.1 Genesis of the concept

The progress that took place in ecological sciences towards the end of the twentieth century resulted in major advancements of understanding and knowledge. A level was attained that enabled an attempt to integrate ecological sciences with more advanced scientific fields, such as physics and hydrology, which together with the application of mathematical tools like modelling created a platform for the development of a new discipline: ecohydrology (Zalewski et al., 1997; Zalewski, 2000).

Ecohydrology is a scientific concept that quantifies and explains relationships between hydrological processes and biotic dynamics at a catchment scale and is applied to solving environmental problems (Zalewski, 2006). It has been defined as a sub-discipline of hydrology focused on ecological aspects of the water cycle. This concept is based on the assumption that the sustainable development of water resources is dependent on the ability to restore and maintain the evolutionarily established processes of water and nutrient circulation and energy flows at the catchment scale.

With respect to the atmospheric/terrestrial and aquatic phases of the water cycle, two independent developments have occurred:

- Progress in understanding the interplay between water-plant-soil (the atmospheric/terrestrial phase of ecohydrology), attained as a consequence of the cooperation of hydrologists, plant physiologists and soil scientists (Eagelson, 1982; 2002; Rodriguez-Iturbe, 2000; Baird and Wilby, 1999), provide a background for the use of landscape management in the regulation of the water cycle.
- Development of the 'aquatic phase' of ecohydrology (Zalewski et al., 1997) is based mostly on the integration of advances in hydrology and the recent progress in limnology. It not only provides the potential to predict the processes in the aquatic environment (e.g., those related to eutrophication) but, based on an understanding of abiotic-biotic interactions, it allows attempts to regulate naturally-established processes of water, nutrient circulation and energy flow in aquatic ecosystems in order to reverse their degradation and enhance the absorbing capacity.

Following this, the key principle of ecohydrology is the 'dual regulation' (Zalewski, 2006) of biota by hydrology (e.g., the regulation of biotic interactions by manipulating reservoir's hydrodynamics for reduction of eutrophication symptoms) and vice versa – hydrology by biota, (e.g., in constructed wetlands; Mitsch and Jorgensen, 1989; 2004) for enhancing the absorbing capacity of ecosystems against impacts. For sound integrated urban water management (IUWM), integration of both the atmospheric/terrestrial and aquatic ecohydrology phases is necessary.

6.2.2 Creating opportunities for the degraded environment

The importance of the ecohydrological approach in the enhancement of an ecosystem's absorbing capacity increased following the publication of a paper by Crutzen and Stoermer (2000) who defines the present era as the 'Anthropocene'. Meybeck (2003) based on an in-depth analysis of literature, demonstrated that the modification of aquatic systems by human pressures (aquatic habitat degradation, flood regulation, ecosystem fragmentation, sedimentation imbalance, salinization, contamination, eutrophication, etc.) has increased to a level that can no longer be controlled by natural processes like climate, relief and vege-tation alone. This is particularly true in highly-modified urban areas, where the level of pressure is so intense that the self-restoration and self-regulation of ecosystems is impossible. The decline of water quality and biodiversity and the increase in flood risks result to a great extent from the degradation of natural cycles. For instance, increased diffuse pollution results from the disruption of biogeochemical cycles (due to unification or fragmentation of landscape, deforestation, biodiversity decrease), while increased flood risk results from a modification of the water cycle (e.g., an increase in impervious surfaces in cities and a decrease in vegetated sections of catchments leading to the intensification of runoff). This provides evidence that the traditional 'mechanistic' approach focused on the elimination of threats, based, for example, exclusively on the reduction of point-source pollution by conventional wastewater treatment plants or flood control by only hydrotechnical solutions, cannot solve the problem. Elements of the technical approach, especially in highly-impacted urban areas, remain valid and necessary; however, they should be complemented by the amplification of environment opportunities, enabling efficient functioning and enhancing their defence mechanisms (see Figure 6.1). In the case of water management practices, that means the extension of the number of potential tools used to reconstruct ecological processes in landscapes and aquatic habitats and to improve their assimilative capacity. Otherwise, water management constitutes a more trial-and-error approach than the implementation of a sound policy towards sustainable water use.

Implementation of the aquatic phase of ecohydrology is to a great extent dependent on the quality of aquatic habitats. As described in Chapter 2, the hydrological features, such as water depth, flow velocity and variability, discharge, flood and drought characteristics, constitute one group of the five major features of aquatic habitats supporting aquatic life. According to the ecohydrology approach, understanding the relationships between hydrological features and biological processes may be a useful management tool. Its application is based on the following tenets:

- dual regulation of hydrology by managing biota and, vice versa, regulation of biota by altering hydrology
- integration of various types of regulation at the catchment scale to achieve synergy serving to stabilize and improve the quality of water resources;
- harmonization of ecohydrological measures with hydrotechnical solutions – for example, dams, irrigation systems, sewage treatment plants, levees in urbanized areas, etc. (Zalewski, 2006).

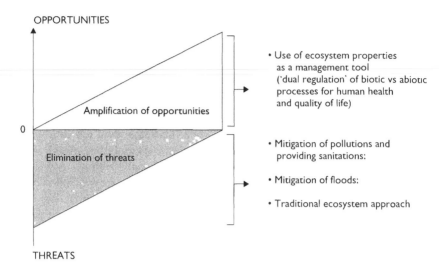

OPPORTUNITIES

Amplification of opportunities

0

Elimination of threats

THREATS

• Use of ecosystem properties
  as a management tool
  ('dual regulation' of biotic vs abiotic
  processes for human health
  and quality of life)

• Mitigation of pollutions and
  providing sanitations:

• Mitigation of floods:

• Traditional ecosystem approach

Figure 6.1 Ecohydrology as a factor maximizing opportunities for aquatic habitat rehabilitation in Integrated Urban Water Management

Source: Zalewski, 2002.

The empirical experience gained over ten years of development and implementation of the ecohydrology concept and its testing in a range of catchments and ecosystems subject to low- or medium-level disturbances include already several examples of research and implementation:

- The relationship between vegetation, soil and water has been elucidated on the basis of the understanding of physiological properties of plants, as presented by Eagelson (1992; 2002) and Baird and Wilby (1999).
- Considerable progress has been made in understanding the role of vegetation in water-cycle processes at the landscape scale, through research by Rodriguez-Iturbe (2000) and that done within the International Geosphere-Biosphere Programme (IGBP) and the Biospheric Aspects of the Hydrological Cycle (BAHC) project (Vörösmarty, 2000).
- The multifaceted role of buffering by ecotone zones between land and water, which have been well defined within the framework of the UNESCO Man and the Biosphere (MAB) Programme (Naiman et al., 1989; Schiemer et al., 1995; Zalewski et al., 2001; Gilbert et al., 1997)
- The application of ecological engineering, such as the management of wetlands for water purification by reducing excessive nutrient loads, on the basis of the ecological theory and mathematical modelling, as developed by Mitsch and Jorgensen (1989)
- The effect of hydrological regimes on vegetation succession of grasslands and swamps, as analysed by Witte and Runhar (2001)
- The reduction of nutrient loads to lowland reservoirs by enhancement of nutrient retention in floodplains, as demonstrated by Wagner and Zalewski (2000)
- The control of eutrophication symptoms and the elimination of toxic algal blooms through the regulation of water levels for control of trophic cascades, as discussed by Zalewski et al. (1990; 2000) and water retention time (Tarczynska et al., 2001)

- The research undertaken on the control of water quality and dissolved oxygen content in ice-covered dam reservoirs, during the winter, by regulating the outlet operation (Timchenko et al., 2000)
- The regulation of the timing of water releases into downstream rivers, in order to maintain fish migration, preserve biodiversity and fish production (the Parana River below the Porto Prima Vera Dam), as investigated by Agostinho et al. (2005)
- The examination of possibilities for managing coastal waters and diminishing their eutrophication using ecohydrology at a catchment scale, as initiated by Wolanski et al. (2005) and Chicharo (2006).

6.3 ECOHYDROLOGY FOR THE URBAN ENVIRONMENT

The empirical experience from medium-disturbed ecosystems should be translated into problem solving approaches in more impacted systems, especially in overpopulated and economically constrained urban areas. As mentioned before, compared to natural and semi-natural systems, the pathways of water, energy and matter in urban spaces are extremely constrained, and their flows are intensive. Prudent control of these flows and their transition through the urban system is crucial for reducing environmental degradation. It is important that proposed measures also be cost efficient in order to make them affordable for the society. Since urban environments are now inhabited by over half of humanity, this issue has become particularly critical for the achievement of the Millennium Development Goals (MDGs) set by the United Nations.

One fundamental assumption in sustainable urban water management is the separation of water resources from nutrients/pollutant cycles (Maksimović, 2001). Highly concentrated streams of pollutants may easily overload the absorbing capacity of the biological system. Once this capacity is exceeded, the resilience of an ecosystem, expressed as the ability to maintain structure and function under stress, dramatically declines (see also Chapter 2). In order to avoid such situations, urban technical infrastructure has to be complemented with constructed or managed ecological systems with a high potential for pollutant and water retention. This functional integration of a city's water services and infrastructure with green areas and their inherent components, as well as aquatic habitats, constitutes an additional challenge. Some efforts towards achieving such integration are given in Chapter 9 and relate to the harmonization of aquatic habitats with stormwater infrastructure (Lodz, Poland and Lyon, France), meeting biodiversity conservation and flood protection (Vienna, Austria) or improving biodiversity, recreational value and quality of life (Istanbul, Turkey and Sharjah, United Arab Emirates).

According to the first principle of ecohydrology, the quantification of the water cycle at the catchment scale should be the starting point for the formulation of a systemic approach (Zalewski, 2000; 2005; 2006). It should be based on analysis of the distribution of aquatic habitats and pollution sources in the city space and analysis of water-mass dynamics in various climatic conditions. Such analyses permit the identification of hot spots, where actions should be undertaken for reducing the probability of flood generation and pollutant discharge, and are the starting point for the separation of water and pollutant (matter) cycles.

Urban aquatic habitats are usually under elevated hydraulic stress, due to high imperviousness in catchment surfaces and the resulting intensification of runoff (see Chapters 2 and 4). Simplified stream channels are converted into raceways for the accelerated transfer of water out of the city space. Such hydrotechnical approaches to flood protection were too often adopted without considering that accelerated outflow from city catchments reduces infiltration and lowers groundwater levels. Such conditions seriously affect the physiological processes of landscape vegetation (e.g., rate of growth, transpiration and pest resistance) and reduce their capacity for water cycle regulation, soil formation and water (and stormwater) retentiveness. The new paradigm has gradually abandoned the general pattern of the rapid removal of excess water from the city landscape and taken into account the landscape potential for its storage and groundwater recharge. Deceleration of runoff during wet weather reduces peak flows without inducing flood risk. It also mitigates dry weather low flows and their impact on water quality and aquatic ecosystem stability. Runoff and stormwater transfer can be reduced by using natural and man-made biotic structures. Promoting diversified vegetation cover and green spaces in city landscapes in order to retain water in the terrestrial phase of the water cycle, is one means. Additionally, the creation of constructed wetlands and impoundments in river valleys, the creation or restoration of floodplain areas, the introduction of detention polders, and the construction of wetlands at stormwater outlets all help to increase water retentiveness in aquatic habitats. Such an approach also reduces the amount of pollutants transferred with surface runoff by enhancing sedimentation and assimilation in vegetation biomass by phytoremediation (see Chapter 5). According to the second principle of ecohydrology, understanding how vegetation adapts to the hydrological characteristics of natural or constructed aquatic habitats may enhance both water retentiveness and pollutant assimilation. An example of such an approach in the City of Lodz, Poland is given in Chapter 9. Results of pilot experiments show that the rate of nutrient and heavy metal uptake by vegetation increases up to twenty times if properly adjusted to hydrological conditions. Such adaptations, based on the understanding of ecosystem properties, can be used as tools in urban aquatic ecosystems to increase their absorbing capacity against pollution.

Several examples of ecohydrological management measures are based on biotic structure response to changes in the hydrological features of habitats. One example is the reduction of man-made reservoirs supplied with nutrients by enhancing their retention on floodplains (Wagner and Zalewski, 2000) and reducing the reservoir eutrophication based on the relationships between patterns of river fluxes and the intensity of toxic algal bloom appearances (Zalewski et al., 2000; Tarczynska et al., 2001). Management of the hydraulic residence time and hydrodynamics of impoundments can be used for sedimentation control and the improvement of self-purification rates. Another example is the reduction of eutrophication symptoms (toxic cyanobacterial blooms) by regulating habitat hydrodynamics and subsequent biotic structure dynamics (Zalewski et al., 1990) and shortening the hydraulic residence time (Tarczynska et al., 2001). These experiments were conducted on the reservoir supplying a city of 800,000 inhabitants in Poland with drinking water.

6.4 MULTIDIMENSIONAL BENEFITS OF THE ECOHYDROLOGICAL APPROACH FOR THE URBAN ENVIRONMENT AND THE SOCIETY

The term 'green city' is synonymous with the notion of a healthy urban environment with a high quality of life. Moreover, it implicitly means that a significant part of the

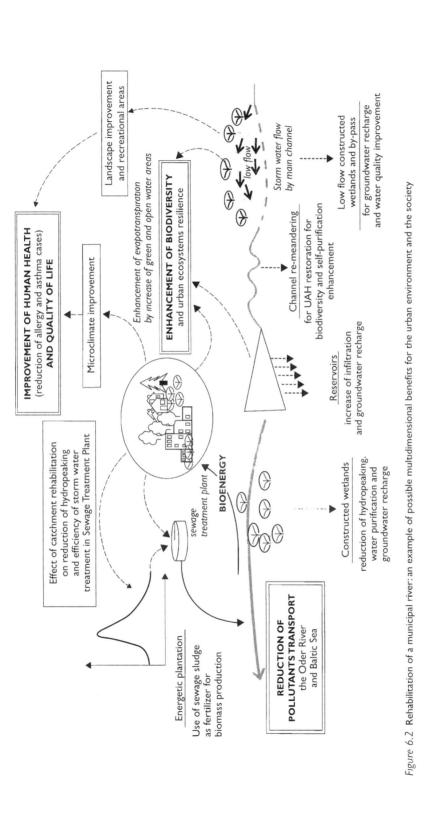

Figure 6.2 Rehabilitation of a municipal river: an example of possible multidimensional benefits for the urban environment and the society

urbanized space is covered by semi-natural terrestrial and aquatic ecosystems. Freshwater and terrestrial ecosystems have an excellent potential for moderation and control of the water cycle and pollution (see Figure 6.2) that should be considered while management plans are being developed. Such urban areas provide citizens not only with regulatory ecosystem services but also aesthetic, cultural and recreational values. However first and foremost, they improve human health in direct and indirect ways (see Chapter 8). There is growing evidence that higher and more stable moisture of city air reduces dust levels, which in turn reduces asthma, allergies and other related diseases. Furthermore, the opportunities for recreation in green areas are important for the proper physical and psychological health of inhabitants.

Well-managed water habitats are visually the most attractive elements of modern cities landscapes, and are usually considered by city planners as 'axes' or 'nuclei,' around which individual green areas and urbanized spaces are functionally organized (see Chapter 7). One of the key questions posed by socio-economists concerning city development constraints in the era of globalization was, what is the major factor for successful city development in Europe? The answer is not surprising: creative and innovative leaders. Following that line, the next question was, what is the primary factor that they consider while choosing a given city as a place to live in and work? The answer was quite surprising: the quality of the environment.

Based on this information, a 'city of the future' should make efforts to expand its attractive living space by restoring and enhancing urban water bodies and aquatic ecosystems as integral and functional elements of the cityscape. An example of such an approach is given in Box 6.1. This example demonstrates a fundamental challenge for the city of the future, based on harmony between the environment, society and the city's primary functions. This harmony can be supported by applying a ecohydrological system approach to the regulation of the interplay between water and biota in the urban space.

6.5 IMPLEMENTATION OF THE ECOHYDROLOGICAL APPROACH

A general scheme of the development and implementation of the ecohydrological approach in the city of the future (see Figure 6.3) has to consider the context of generally accepted goals like sustainable development (see Chapter 3) and documents creating a vision for environmental problem-solving (e.g., the United Nations Millennium Declaration). From a strategic point of view, it should be compatible with already implemented policies, such as IUWM or the EU Water Framework Directive.

The first step should include an integrative analysis of the hydrological and ecological dynamics of the catchment, as well as an assessment of the quality of aquatic habitats and ecosystems, e.g., on the basis of habitat features or a bioassessment (see Chapters 2 and 4). At this point, adoption of the principles of ecohydrology is necessary for developing an implementation plan.

The next step involves an integrative analysis of the dynamics of hydrological and biological processes. The existence of parallel data on biological system processes under different hydrological conditions allows for assessing the ecohydrological dynamics of the system and regulatory feedbacks for potential application in water management schemes (e.g., the rate of self-purification at different temperatures and stages of the hydroperiod). Identification of the range of possible regulatory feedbacks between ecological

BOX 6.1 Ecohydrology for strategic city planning: City of Lodz, Poland

The case of Lodz provides an example of ecohydrology application benefits for both ecology and society. The most attractive residential area of the city (Julianów) is connected to the southern part of a large forest area with a nature reserve and several impoundments on headwater streams of the Bzura River. The valley of a small urban stream (Sokolowka), with numerous patches of natural vegetation and agricultural land has been neglected and considered a remote area (see the figure below). The creation of a natural extension of this area by constructing impoundments and recreation facilities (including cycle paths, jogging routes, picnics areas, etc.) has attracted the interest of developers. Now the Sokolowka River Valley is under restoration, which has made it more attractive for development. This ongoing project aims at improving the cultural and aesthetic values of the landscape, creates employment opportunities, increases the city's taxation income, enhances the quality of life, reduces air pollution, improves the microclimate and, consequently, the health of the urban population.

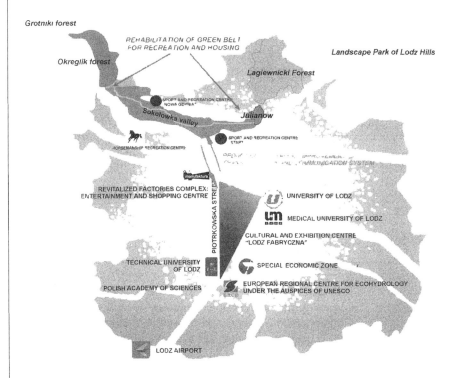

The figure above presents the concept of rehabilitation in the Sokolowka River Valley for the creation of a recreational space serving the city inhabitants. (See also colour plate 6) The area is located close to the city centre, which has become known as the 'Triangle of Dynamic Development' due to recent intensive development. The location also makes it an attractive residential area for developers, leaders and employees, providing employment opportunities for the workforce at all levels and qualifications, which in turn contributes to the improvement of the city's economic and social situation.

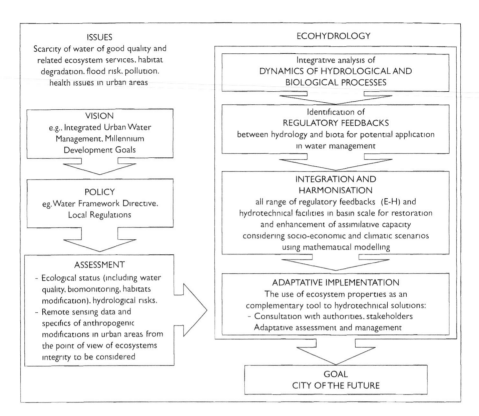

Figure 6.3 Schematic of implementation of the ecohydrological approach in urban areas

Source: Zalewski, 2004.

and hydrological processes and hydrotechnical facilities in the catchment is an important factor for aquatic habitat management and the rehabili-tation and enhancement of the system's assimilative capacity.

All of the above measures have to be developed together with a consideration of socio-economic conditions. The prerequisite of the implementation of ecohydrology is public involvement, which encourages decision-makers and the public to accept and follow the new way of thinking. Until now, education and stakeholder discussions have been considered among the best ways. However, comparative research on several ecohydrology projects seems to imply that a good starting point is the opportunistic character of human nature, seeking a demonstration of the tangible benefits, such as threat reduction (e.g., floods) and the enhancement of ecosystem services (e.g., water quality improvement, biomass/bioenergy production and biodiversity). In many societies, especially those in countries of limited resources and low economic status, environmental conservation is considered a luxury, or at least a target with questionable value to the society. This is why it is important to demonstrate that environmental conservation may generate ecosystem services and, as a consequence, employment, which then amplifies a cascade of positive socioeconomic and environmental feedbacks. The experience gained by the UNESCO and UNEP-IETC Demonstration Project in the Pilica River Valley (Wagner-Lotkowska et al.,

2004) has contributed to the conclusion that, especially in the cases of increasing tensions between society and water/environmental resources, the most efficient stimulation for public involvement should the assumption of 'ecosystem services first' (Zalewski, 2004). A visible and profitable outcome for society should be provided in the first stage of implementation. This is why adaptive assessment management, taking into account possible drivers and adapting to them, becomes the best way not only for understanding the complexity of ecohydrology feedbacks and revising the new management strategy, but also for further implementation progress on the basis of social acceptance.

One of the social benefits is the reduction of health risks in densely populated city areas, which is discussed in Chapter 8.

ACKNOWLEDGEMENTS

The concept presented here was developed by the European Regional Centre for Ecohydrology under the auspices of UNESCO, the Polish Academy of Sciences and the Department of Applied Ecology of the University of Lodz. It was possible thanks to a long-term cooperation with the City of Lodz Office, in particular the Department of Infrastructure. Lodz is a Demonstration City in the 6th European Union Framework Program Integrated Project 'SWITCH' (GOCE 018530).

REFERENCES

Agostinho, A.A., Gomes, L.C, Verissimo, S. and Okada, E.K. 2005. Flood regime, dam regulation and fish in the Upper Parana River: effects on assemblage attributes, reproduction and recruitment. Reviews in *Fish Biology and Fisheries* (2004) 14: pp. 11–19

Baird, A.J. and Wilby, R.L. (eds). 1999. *Eco-hydrology. Plants and water in terrestrial and aquatic environments.* Routledge, London, New York.

Chicharo, L. and Chicharo, M.A. 2006. Applying the ecohydrology approach to the Guadiana estuary and coastal areas: Lessons learned from dam impacted ecosystems. Editorial. Estuarine, *Coastal and Shelf Science* 70, 1–2.

Crutzen P. J. and Stoermer E. F. 2000. The 'Anthropocene'. *IGBP Newsletter.* No 41.

Eagelson, P.S. 1982. Ecological optimality in water limited natural soil-vegetation systems. 1. Theory and hypothesis. *Water Resource Research.* 18: pp. 325–40.

Eagelson, P.S. 2002. *Ecohydrology.* Cambridge Scientific.

Gibert, J., Mathieu, J. and Fournier, F. (eds). 1997.Groundwater / Surface Water Ecotones: *Biological and Hydrological Interactions and Management Options* UNESCO, International Hydrology Series, Cambridge University Press.

Maksimovič, C. (ed.) 2001. Urban Drainage in Specific Climates. IHP-V Technical Documents in Hydrology. No 40, Vol. III, M. Nouh (ed.) Urban drainage in arid and semi-arid climates. UNESCO Paris.

Meybeck, M. 2003. Global analysis of river systems: from Earth system controls to Anthropocene syndromes. *Philosophical Transactions of the Royal Society of London [B]* 358(1440), pp. 1935–55.

Mitsch, W.J. and Jorgensen, S.E. (eds). 1989. *Ecological engineering. An introduction to ecotechnology.* John Wiley, New York, Chichester, Brisbane, Toronto, Singapore.

Mitsch, W. and Jørgensen, S.E. 2004. *Ecological Engineering and Ecosystem Restoration,* John Wiley and Sons, US.

Naiman, R.J. and Decamps, H. 1990. *The ecology and management of aquatic-terrestrial ecotones.* UNESCO MAB series. Parthenon / UNESCO Paris, France.

Rodrigues-Iturbe, I. 2000 Ecohydrology: A hydrological perspective of climate-soil-vegetation dynamics. *Water Resource Research*, 36:pp. 3–9.

Schiemer, F., Zalewski, M. and Thorpe, J. (eds). 1995. The Importance of Aquatic-Terrestrial Ecotones for Freshwater Fish. *Developments in Hydrobiology* 105. Kluwer Academic Publishers.

Tarczynska, M., Izydorczyk, K. and Zalewski, M. 2001. Optimization of monitoring strategy for eutrophic reservoirs with toxic cyanobacterial blooms. Proceedings of 9th International Conference on the Conservation and Management of Lakes, Otsu, Japan. 3C/D–P83: 572–575.

Timchenko, V., Oksiyuk, O. and Gore, J. 2000. A model for ecosystem condition and water quality management in the Dniepr River Delta. *Ecol. Eng.* 16, pp. 119–25.

Vörösmarty, C.J. and Sahagian, D. 2000. Anthropogenic disturbance of the terrestrial water cycle. *Bioscience* 50: pp. 753–65.

Wagner, I. and Zalewski, M. 2000. Effect of hydrological patterns of tributaries on biotic processes in a lowland reservoir – consequences for restoration. *Ecological Engineering* 16: pp. 79–90.

Wagner-Lotkowska, I. Bocian, J., Pypaert, P., Santiago-Fandino, V., Zalewski, M. 2004. Environment and economy – dual benefit of ecohydrology and phytotechnology in water resources management: Pilica River Demonstration Project under the auspices of UNESCO and UNEP. *Ecohydrology & Hydrobiology Journal*. Special Issue: Ecohydrology from Theory to Action. No 3, pp. 345–52

Witte, J.P.M., Runhar, J. and van Ek, R. 2001. Three approaches to ecohydrological modelling. International Hydrological Programme, Ecohydrology – from concepts to applications. Cambridge University Press, Cambridge.

Wolanski, E., Boorman, L.A., Chicharo, L., Langlois-Saliou, E., Lara, R., Plater, A.J., Uncles, R.J. and Zalewski, M. 2005. Ecohydrology as a new tool for sustainable management of estuaries and coastal waters. *Wetlands Ecology and Management* 12(4), pp. 235–76

Zalewski, M. 2000. Ecohydrology. The scientific background to use ecosystem properties as management tools toward sustainability of water resources. Guest Editorial, *Ecological Engineering*, 16: pp. 1–8.

Zalewski, M. (ed.) 2002. Guidelines for the Integrated Management of the Watershed. Phytotechnology and Ecohydrology. UNEP, Division of Technology, Industry and Economics. *Freshwater Management Series* No.5.

Zalewski, M. 2004. Ecohydrology as a system approach for sustainable water biodiversity and ecosystem services. *International Journal of Ecohydrology and Hydrobiology*. Proceedings of the International Conference 'Ecohydrology from Theory to Action' 18–21 May 2003, Wierzba, Poland, in the framework of the UNESCO MAB/IHP programmes, Vol 4, No 3, pp. 229–36.

Zalewski, M. 2005. Engineering harmony. *Academia*, 1(5): pp. 4–7.

Zalewski, M. 2006. Ecohydrology – an interdisciplinary tool for integrated protection and management of water bodies. *Arch. Hydrobiol. Suppl.* 158/4, pp. 613–22

Zalewski, M., Tarczyńska, M. and Wagner-Lotkowska, I. 2000. Ecohydrological approaches to the elimination of toxic algal blooms in a lowland reservoir. *Verh. Internat. Verein. Limnol.* 27, pp. 3178–83.

Zalewski M., Thorpe J. and Naiman R. 2001. Fish and riparian ecotones – a hypothesis. *Ecohydrol. Hydrobiol.* 1:pp. 11–24.

Zalewski, M. and Naiman, R.J. 1985. The regulation of riverine fish communities by a continuum of abiotic-biotic factors. *Habitat Modifications and Freshwater Fisheries* J.S. Alabaster (ed.), FAO, Butterworths, London, pp. 3–9.

Zalewski, M., Brewinska-Zaras, B., Frankiewicz, P. and Kalinowski, S.1990. The potential for biomanipulation using fry communities in a lowland reservoir: cocordance between water quality and optimal recruitment. *Hydrobiologia*, 200/201: pp. 549–56.

Zalewski, M., Janauer, G.A. and Jolankai, G. 1997. Ecohydrology. A new paradigm for the sustainable use of aquatic resources. UNESCO IHP Technical Document in Hydrology No. 7; IHP – V Projects 2.3/2.4, UNESCO Paris.

Chapter 7

Integrating aquatic habitat management into urban planning

Elizabeth DAY[1], Giovanna BRAIONI[2] and Azime TEZER[3]

[1] Freshwater Research Unit, University of Cape Town, Rondebosch, 7700, South Africa
[2] Department of Biology, Padova University, Via U.Bassi 58/B, 35121 Padova, Italy
[3] Istanbul Technical University, Faculty of Architecture, Urban and Regional Planning Department, Taskisla 34437 Taksim/Istanbul, Turkey

7.1 BIODIVERSITY AND THE EMERGENCE OF SUSTAINABLE DEVELOPMENT PRACTICES IN URBAN LANDSCAPE PLANNING

Despite widely differing cultural, social and economic differences across the globe, the impact of urban development on aquatic habitats has been remarkably similar throughout the spectrum of developing and developed nations (Gopal and Wetzel, 1995; 1999; 2001; 2004). Several examples of such issues, arising in urban areas located all over the world – from Africa to North America and the Arabian Peninsula to Europe – are presented in Chapter 9 of this book. Common issues include poor water quality, a loss of biodiversity, habitat degradation, poor ecological functioning and a loss of ecological, hydrological connectivity. They also include a loss of social connectivity and conflicts between the needs and concerns of different user groups, including local authorities, private and business sectors, environmental managers, conservation agencies, human health and welfare agencies and the aquatic ecosystems themselves (Day et al., 2005). Such problems are usually rooted in what has historically been a tightly focused single-disciplinary approach to various management issues in an urban context, without consideration of the complexity of interactions between the built environment and ecological, environmental, hydrological, and geohydrological issues. These in turn interact with a complex matrix of economic and social objectives and limitations, at the local, catchment and landscape scales (Braioni et al., 2006a; 2006b). Even where efforts have been made to integrate natural resource management and land-use planning, the final outcomes are often compromised by control at administrative and implementation levels by an array of different sectors and levels of government (Conway and Lathrop, 2005).

In recognition of these problems, planning and urban design professionals began to investigate means of establishing better interactions between natural ecosystems and urban development, not only at the level of urban aquatic habitats but also with regard to the need for incorporating natural ecosystems into the built environment at a landscape level (Tezer, 2005). As early as the start of the 1900s, movements such as The City Beautiful Movement and The Garden City Movement focused on re-introducing and/or incorporating natural systems into the fabric of urban landscapes (Levy, 2000), and concepts such as McHarg's (1969) Design with Nature had a significant influence on the development of ecologically sensitive urban planning practices in the latter part of the twentieth century (Tezer, 2005).

By the early 1990s, increasing attention was being paid to sustainable development policies. These aimed at balancing interactions between environmental, economic and social dimensions in rapidly changing urban areas and included the classification of the United Nation's Commission on Sustainable Development's Theme Indicator Framework (DiSano, 1995; Beatley and Manning, 1997; OECD, 2002; Portney, 2003; UN-CSD, 2005) and the Millennium Ecosystem Assessment Report, Biodiversity and Human Well-being: A Synthesis Report for the Convention on Biological Diversity. Such programmes recognize the urgency of implementing practices such as eco-sensitive urban transformation, habitat rehabilitation, environmental awareness, stewardship activities, eco-jobs, and integrated watershed planning, monitoring and biodiversity protection in urban areas, if development is to be sustainable. Critical to this is the need for an integration of biodiversity and urban planning practices (Sukopp and Weiler, 1988; Gyllin, 1999; Hahn, 2002; Alfsen-Norodom et al., 2004; Tezer, 2005).

Integrating biodiversity issues into urban areas becomes more complex when it is the biodiversity of aquatic ecosystems that is under consideration. This is because both water and biodiversity are cross-sectoral issues, with the former affecting all levels and activities of ecosystems and society and the latter spanning several different environmental and social sectors. Since water policy is cross-sectoral in character, (Roux et al., 2006) it follows that biodiversity policy should be similarly cross-sectoral. Integrating the resulting complexity of inter-related policies is thus highly challenging. Where the biodiversity of aquatic ecosystems in urban areas is the focus, it is critical to ensure that, at the policy level, both respective sets of policies and management instruments are at once coherent and aligned with each other in order to avoid conflicting objectives and contradictory management approaches (Roux et al., 2006).

7.2 THE NEED FOR INVENTORIES OF AQUATIC HABITATS

There are three important questions related to aquatic habitats in the context of the urban environment: 1) the extent to which competing demands on aquatic ecosystems can be met within the urban environment, 2) the level to which different management objectives impact on natural ecosystem functioning and 3) the degree to which this is of concern within an urban context. These questions can only be answered with accurate and detailed information regarding the type of aquatic ecosystems that are present both in the urban area and within the greater watershed or catchment area, their ecological condition, current trajectory, vulnerability to particular types of impact and, critically, the importance of the components of a particular ecosystem in a local and a broader regional, or even global, context.

In most urban scenarios, this last point has little relevance. Wade et al. (1998) make the point that urban aquatic habitats usually have only local value in terms of their contribution to conservation, with less-impacted habitats usually occurring in rural, less disturbed environments. This holds true for many systems, particularly in ecoregions where natural endemism is low and even unimpacted species diversity exhibits little change on the scale of local and sometimes regional landscapes. By contrast, urban freshwater ecosystems in some ecoregions may exhibit very high endemism. Given these quite different scenarios, it follows that effective conservation and management of urban ecosystems needs to take place against a background of adequate data, including spatial information regarding an ecosystem's location, type, function, present condition and, where possible, reference condition (Finlayson and van der Falk, 1995; Dini and Cowan, 2000). Case studies such as the Wasit wetland reserve study (see Chapter 9) highlight the importance of carrying out adequate baseline assessments before embarking on intervention projects where there is a poor understanding of the present trajectory or driving variables of a system. The importance of monitoring and defining the ecological status of the water bodies for identifying water quality objectives is illustrated in the example of the Adige River Fluvial Corridor Case Study in Italy. By contrast, the research in the Yzeron Catchment in Lyon, France highlights the issue of the identification of differential capacities for pollution assimilation along different reaches of an urban watercourse. Urban stormwater planning takes into consideration patchiness in stream ecosystem resilience and functional capacity in designing discharge points.

Addressing the above concerns and allowing effective strategic incorporation of aquatic habitat into a development framework assumes that a detailed inventory of aquatic ecosystems likely to be affected by activities associated with the urban environment is available. Such an inventory should provide accurate spatial information regarding the distribution, hydrological and geohydrological linkages of affected freshwater ecosystems, their resilience absorbing capacity to different kinds of impacts associated with the urban environment, the degree to which each has been altered from its natural condition and its relative conservation importance. Furthermore, a strategy for assigning long-term management objectives for different systems, based on meeting the goals of various social, ecological, hydrological or other criteria, should also be available. Such an approach enables contributions of future developments to long-term and positive ecosystem management, rather than to the gradual accumulation of permanent and unmitigable impacts leading to a cycle of degradation and a loss of biodiversity and basic ecosystem function.

In practice, however (particularly in developing countries) the status and often even the existence of many wetland and riverine ecosystems is often poorly understood. Many of them are only recognized once expanding development has set in place immovable impacts and constraints on future rehabilitation. In areas where biodiversity stakes are high, the implications of this may be permanent loss (extinction) of species or even whole communities. Examples of this occur in the aquatic habitats of Istanbul and the Cape Floral Kingdom of South Africa, where Cape Town and Istanbul are examples of urban settings in areas of high biodiversity abundance and low biodiversity resilience (Day et al., 2005).

Thus compilation of a wetland and river inventory, including spatial, biological and hydrological data, as well as any other relevant and available information, should be an integral early informant of urban planning, and should include categorization of river and wetland type at a level that allows overlay of management and rehabilitation information between systems of similar type and impactedness (Bostelmann et al., 1998). The Procedural/Actions Abacus (Braioni and Braioni, 2002) presented later in this book is one example of a tool developed to allow this level of integration (the follow-up of this section is presented in Chapter 9).

7.3 INCORPORATING WETLANDS AND RIVERS INTO URBAN PLANNING AND LAYOUT

Fundamental to the sustainability of aquatic habitats (rivers and wetlands) in urban environments is the need to incorporate them into the urban framework on a strategic level. This facilitates their ecological and social functions in terms of:

- their contribution to urban open space and the recreational and amenity needs of urban communities: the case study of the Lobau Biosphere Reserve in Vienna, Austria (see Chapter 9) provides such an example. Saving high ecological values of a wetland and floodplain of the Danube River Corridor provides areas of recreation and environmental education, while combining the system with flood protection for the city;
- their effectiveness as ecological corridors through otherwise ecologically hostile urban areas: the width of corridor needs to take into consideration the kinds of ecological functions that need to be maintained within the corridor – this aspect is explored further in the discussion around the establishment of ecological buffer areas;
- their capacity to provide linkages between nodes of important terrestrial and /or aquatic habitats, for example, effecting linkages between mountain and coastal ecosystems. Such linkages are not only important to aquatic species with known migratory requirements in their lifecycles, but also for preventing the fragmentation of different natural communities and ongoing genetic isolation, as well as in providing opportunities for recolonizing previously impacted habitats;
- their ability to provide protection of downstream ecosystems that may be of high conservation importance or sensitivity: the strategic management of different freshwater ecosystems within the urban landscape for specific functions can be an effective way of improving long-term overall ecosystem health. Such is the case, for example, in the sacrificial use of certain degraded wetland habitats, or portions of wetlands, for water quality improvement, with the positive secondary effect that more sensitive downstream habitats may thus be protected from poor water quality impacts and managed to support more sensitive or ecologically important habitat types or quality.

In addition to the provision and maintenance of natural habitats, aquatic ecosystems are increasingly being recognized by municipalities and local governments in urban areas for their role in providing a wide range of useful goods and services. These range from flood attenuation (see Chapter 9) and erosion control, to storm water conveyance and management (see Chapters 4 and 9), water quality amelioration and the provision of amenity and recreational services and urban greenbelts. However, it is also recognized that the biological health of aquatic communities in urban areas can be negatively correlated with the amount of urban land-use in the surrounding watershed (Miltner et al., 2004).

As demands for space increase and densification of urban areas becomes the norm, increasing pressure is placed on aquatic systems to provide multiple services, among which there may be little complementarity. Day et al. (2005) note that there is often conflict between competing objectives in the management of riverine corridors and wetlands, many of which are mutually exclusive. Thus the recognition of the potential for aquatic systems to provide goods and services must be tempered by the reality that aquatic systems do not all provide the whole range of goods and services attributed to these ecosystems in general.

7.4 THE ROLE OF AQUATIC HABITATS IN FACILITATING URBAN BIODIVERSITY CONSERVATION

One of the results of increasing pressures on urban areas is the fact that riverine corridors are increasingly recognized for their potential (and essential) contribution to conservation management at a landscape scale (Wade et al., 1998). Often comprising the only area of relatively undeveloped open space in urban and peri-urban areas, riverine corridors can facilitate linkage of remnant patches of natural or otherwise important terrestrial and/or wetland habitat and the potential development of longitudinal and lateral corridors for the movement of animals through otherwise hostile urban environments (Castelle et al., 1992; Wade et al., 1998). Identifying key geo-morphological, hydrological and biological processes and their spatial location with reference to riverine corridors, wetlands and the urban matrix is a critical component in the development of effective spatial planning within urban areas, at a scale that incorporates ecologically meaningful processes and allows long-term sustainability of ecosystem function.

Crucial to the identification of key processes is the scale at which urban planning incorporates aquatic habitats into long-term planning. Impacts on aquatic habitats typically take place at a catchment level (Miltner et al., 2004) and secondary effects from urban systems may be similarly far-reaching. Lee and Dinar (1995) cite fragmented approaches to river basin management as key reasons underlying failed rehabilitation outcomes and inefficient or inappropriate uses of aquatic resources. Thus the identification of important processes that drive aquatic ecosystem characteristics must include an assessment of the scale at which these drivers are important and a highlighting of particular patterns or processes that may require management on a broader scale, sometimes broader than the urban environment itself. In the case study of the Wasit wetlands in the United Arab Emirates, for example (see Chapter 9), key processes determining ecosystem function included connectivity with groundwater source areas and thus management at a catchment level of abstraction and impacts on subsurface water quality. Although these processes may not necessarily fall within the management area of a city or local authority, recognizing their importance as ecosystem drivers is important in establishing the long-term trajectory of change in a system, its vulnerability to impact and the efficacy of intervention measures without recourse to control of major ecosystem drivers.

The reverse side of this also holds true: some activities within urban areas may affect important downstream systems or processes, with long-term biodiversity or functional impacts on a more regional scale. Urban planning processes need to take into consideration such factors. In this regard, although the focus of the present chapter is on spatial planning, another important aspect of the management and conservation of aquatic ecosystems is the introduction or management of disturbance regimes, often critical in maintaining long-term community composition. The long-term sustainability of ecosystems that depend on even occasional disturbance, such as flood scour or deposition, fire and grazing, are all at risk in urban settings unless innovative strategies can be introduced into the urban fabric.

In the case of European Economic Community (EEC) signatory countries, various directives have been developed with stringent requirements that aim to protect natural ecosystems, including aquatic habitats in urban areas, by controlling the allowable impact of a development on the environment, at scales ranging from site-specific measures (contained in Council Directive 92/43/EEC, Article 6, Paragraphs 3 and 4) to broader strategic measures (Council Directive 85/337/EEC), as well as the Water Framework Directive and measures aimed at affording protection at the catchment/basin level.

The extent to which existing aquatic habitats can be conserved, maintained or rehabilitated to some approximation of their natural condition may also be limited in urban areas (see Chapter 2). Major limitations include the difficulty in isolating a section of river or wetland from upstream or catchment impacts, the permanence of existing impacts (e.g. urban runoff, altered flood lines, catchment level impacts, etc.), a lack of pre-disturbance data to guide the restoration process by setting templates of habitat type and finally, social and political issues, which often revolve around human health, safety and security (Jasperse, 1998; Wade et al., 1998). Social and ecological objectives can also be, and often are, at variance with each other. Day et al. (2005) note that in some urban areas, where space is a critical resource and where social upliftment is a priority of local and national government, ecological objectives are often subjugated to objectives such as the provision of recreational areas, or avoiding the creation of areas that provide or could provide opportunities for criminal activities. The Moddergat River case study (see Chapter 9) provides an example of how these conflicting objectives can affect opportunities for river rehabilitation, as does the case study of the Danube River (Lobau Biosphere Project), albeit at a different socio-economic level.

The potential for habitat and biodiversity issues to be included in a project, or for the incorporation of bioremediation measures like wetland filtration systems, is often compromised by these alternative objectives. Nevertheless, most remediation projects, even where driven by social or engineering objectives, have the potential to improve many kinds of urban aquatic habitats if approached from a multi-disciplinary perspective in the design phase. This can be done by increasing spatial, fluvial and temporal diversity within the system or within sections of a system (Day et al., 2005). Clearly, however, the availability of space in the urban framework is a critical factor in determining the quality and diversity of aquatic habitat that can be achieved within fluvial corridors and wetland nodes.

In the city of Cape Town, where conservation of globally significant biodiversity is a high priority, a Biodiversity Network has been developed (Laros, 2004) with the primary objective of conserving both patterns and processes of the city's biodiversity. The approach taken has similarities to the 'Greenways' approach adopted elsewhere (Peck, 1998; Wade et al., 1998), where the conservation of biodiversity and the maintenance of ecosystem goods and services are integrated into open space planning in a manner that makes them a key structuring element of the urban form (Laros, 2004).

According to the Biodiversity Network, all biodiversity areas are categorized in the following way:

- Category A areas should be protected as key biodiversity areas and will include existing nature reserves and other areas that meet the established criteria. They are to be managed primarily for the maintenance of biodiversity.
- Category B areas are managed to maintain biodiversity but also to support other appropriate activities and land-use types.
- Category C areas fulfil a primary function of supporting an activity or land-use that is mutually inclusive of the maintenance of biodiversity but is not managed primarily as a biodiversity area.

Approaches like the Biodiversity Strategy and various versions of the Greenways concept provide useful backdrops against which long-term land-use planning can take place, allowing linkages between habitats in an ecologically fragmented landscape (Wade et al., 1998; Laros, 2004). They are, however, only as strong as the level to which they can be implemented. In the Cape Town scenario, social priorities including addressing critical housing shortages, gangsterism and crime mean that decreasing financial resources are available for conservation through city budgets, making the possibility of formally purchasing areas identified within the biodiversity network unlikely. Nevertheless, identifying a biodiversity network does allow the prioritization of conservation efforts, allowing limited resources to be directed towards areas with the highest biodiversity conservation priority (Laros, 2004). At the same time, through the development of supporting land-use guidelines, the biodiversity network can also direct potential investment within the city.

Other metropolitan areas that have developed strategies that demonstrate the importance of densely urbanized areas in terms of their biodiversity include Berlin, Chicago, Istanbul (see Chapter 9), Rome, São Paulo and Stockholm (Zerbe et al., 2003; Matsuura, 2005; Tezer, 2005; Platt, 2006).

In all cases, a multi-participatory organizational framework, fostering and coordinating biodiversity protection, is necessary for the continuity of these programmes. Designing cities compatible with their ecological context will have a considerable and positive impact on the establishment of sustainable landscapes that are responsive to human needs and foster human understanding of natural history and the protection of biodiversity and global sustainability (Hahn, 2002; Ahern and Boughton, 1994). Urban planning practices thus need to be integrated with ecosystem functioning and human interventions in a multi-dimensional perspective.

The application of the Greenways concept provides for the spatial incorporation of fluvial corridors in urban planning. Some of the previous sections have outlined the critical need for the inventory and categorization of aquatic habitats to inform prioritization within an urban landscape. Key components of categorization should include indicators of wetland or river sensitivity and resilience, hydrological function and the extent to which it provides a habitat for red data species. For countries within the EEC, the European Environment Agency (EEA) has listed a set of measurable biodiversity indicators, against which changes in various aspects of biodiversity can be assessed (EEA, 2003). Similar indices are being, or have already been, developed in many other countries that are signatories on the Convention on Biological Diversity (CBD, 1992).

The EEA indicators facilitate the assessment of the importance of biodiversity in urban and/or rural areas, as well the measurement of the success of efforts towards biodiversity protection. The report identifies twelve sectors, namely nature protection, forestry, energy, recreation/tourism, climate change, urban development, rural development, water, infrastructure/transport, trade, fisheries and agriculture. According to the report, almost 60% of the prescribed indicators are related to the protection of natural resources, ranging from species characteristics, soil characteristics and land management, to criteria for assessing ecosystem risk in relation to human intervention. The document provides a functional framework for the assessment of overall biodiversity characteristics.

A general framework for biodiversity indicators, based on a subset of the EEA (2003) indicators, which vary from economic through social and natural components of urban settlements is provided in Table 7.1 (Tezer, 2005).

Table 7.1 Indicators of Biodiversity Protection Practices in Urban Areas (Tezer 2005)

INSTITUTIONAL

Sub-Theme	Indicators
Municipal Budget	Networking of open spaces and local parks % of land maintained as natural urban landscape by local government Subsidies of local government to community for green space development Number of ecological rehabilitation and/or "urban greening" projects implemented Number of regulations/ordinances supporting biodiversity protection practices Annual municipal expenditures on parks and open spaces
Initiatives, NGOs	Number of initiatives/NGOs related with biodiversity conservation Number of urban stewardships for biodiversity conservation

ECONOMIC

Sub-Theme	Indicators
Economy and Economic Development	% of employed people who practice to improve biodiversity protection in the city Number of economic activities which help biodiversity protection and poverty alleviation

ENVIRONMENTAL

Sub-Theme	Indicators
Environment	Less environmentally destructive management practices (e.g. Eco-tourism, eco-agriculture) Green ways/green corridors design programs Number of schools that integrate environmental education in their curricula
Biodiversity	Programmes for managing, monitoring and protecting biodiversity % of area for habitat restoration in degraded natural environments Removing invasive alien species programmes Abundance of species and habitats counts Quality and quantity of wetlands Training programmes and awareness building for sustainable use of biological sources
Water	% of habitats restored for drinking water quality Number of aquatic habitats protected for water quality
Climate Change & Ozone Depletion	Programmes for battling heat islands' effect of urbanization with green design applications % of trees and vegetation on streets and open parking lots % of green roof programmes
Parks and Open Spaces	% of population with potential recreational facility within a ten-minute walk Number of green street corridors created annually % of parks and open spaces developed/enhanced/restored annually Number of volunteer hours spent annually for improving open spaces and biodiversity protection

Table 7.1 (Continued)

Food and Agriculture	% of local and regional agricultural products Number of public agricultural gardens Programs for sustainable agricultural practices and training at community and school children levels

SOCIAL

Sub-Theme	Indicators
Environmental Justice and Equity	Poverty alleviation programmes related with environmental and biodiversity protection Participation of disadvantaged communities in decision making processes
Human Health	Number of activities organized locally for improving environmental awareness Number of recreational facilities to improve physical health of citizens
Hazard Mitigation and Community Resilience	% of area with high natural disaster risk, planned sensitively for disaster mitigation Number of habitat restoration programmes for disaster mitigation efforts

Source: Outline of the table is based on the UN Commission on Sustainable Development, Theme Indicator Framework, UN-CSD 2005.

7.5 TRADING ECOSYSTEM INTEGRITY WITH PROVISION OF GOODS AND SERVICES IN AN URBAN CONTEXT

In most urban areas, the long-term degradation of habitat means that urban areas usually have only low importance with regard to their capacity to provide a habitat for threatened species in the Red Data Book (Wade et al., 1998). From a strategic perspective, the categorization of aquatic habitats should be a crucial aspect informing the extent to which different systems lend themselves to different functions within the urban environment. For example, the use of wetlands in stormwater management has been recognized for some time, with wetlands playing a role in flood attenuation, water quality amelioration and erosion and sediment control (see Chapter 5). However, implicit in the deployment of wetlands for the provision of such goods and services is the fact that not all wetland types lend themselves to these functions, and their use in such aspects of stormwater management may entail fundamental alterations of their natural, or present-day, characteristics.

The extent to which such changes are of concern from a biodiversity or other perspective should be informed by wetland classification and inventory, with the latter providing informants of ecological sensitivity. As examples of this, the use of naturally seasonal wetlands or rivers in the conveyance pathways of treated sewage effluent and/or stormwater can lead to fundamental changes in ecosystem function and structure, resulting in changes in vegetation type and a change in the kinds of aquatic species that occur there. These changes often favour invasion by exotic fish and other flora and fauna (Day et al., 2005). Where this occurs at the expense of taxa that have been identified as of biodiversity importance in their own right (e.g. red data species or species that have broader ecological importance, such as in natural pollination or dispersal mechanisms), it is important that such tradeoffs should be red-flagged for a careful weighing of ecological versus engineering, economic or social tradeoffs.

One approach to the resolution of such a conflict entails the delineation of urban areas into different management units. Inner city aquatic habitats can be managed primarily for the provision of goods and services with low spatial requirements, and within a context of sustaining low levels of basic ecological function only. Aquatic habitats in peri-urban to suburban areas can be seen at the same time as lending themselves to a far wider array of possibilities in terms of the provision of different aquatic habitat types. The Lodz case study (see Chapter 9) highlights, for example, the use of peri-urban open space for the creation of vegetated channels and broad open water habitats, although in the Lodz case, these were designed primarily from the perspective of provision of amenity opportunities, rather than with direct ecosystem rehabilitation in mind. Such pragmatic approaches allow conservation efforts to be directed to areas where spatial pressures are often reduced and social requirements for open space and recreational areas may not be in conflict.

The above approach is not, however, a generic solution to the management of aquatic habitats in urban areas, but rather depends heavily on the kinds of systems involved. In Cape Town, for example, past fragmentation and the loss of highly restricted habitat types of global biodiversity significance mean that certain habitat types are represented only by fragments in the heart of the city, where development pressures are immense. The wetland heaths *Passerina paludosa* and *Erica verticillata*, for example, are both highly endangered plant species – the former restricted to some 500 plants remaining in the wild, with most habitats occurring on the Cape Flats, Cape Town's most rapidly developing area, with the highest levels of poverty, crime and unemployment, while *E. verticillata* is restricted to one small patch of extant wetland in an urban nature reserve. In this scenario, the management of aquatic habitat needs to take into account isolated habitat fragments. The allocation of broad areas of urban space to high-impact use of aquatic systems (e.g. stormwater detention, water quality treatment, etc.) is inappropriate from a biodiversity perspective, where remnant fragments of important habitat types still exist and should be supported by appropriate land-use and water management strategies. Miltner et al. (2004) caution against the 'shifting baseline syndrome' (Pauly, 1995) in the management of valuable ecosystem resources, whereby the gradual loss of ecosystem goods and services goes unnoticed or is even culturally normative in urban aquatic systems as a result of the slow but cumulative effects of suburbanization.

Another approach is, by contrast, the allocation of scarce resources for the rehabilitation of degraded systems to a level of improved, but probably not natural, ecosystem functioning in areas where biodiversity conservation is not a primary concern. This approach, it might be argued, should focus on areas in the urban framework where existing developments have not yet resulted in permanent constraints on the width of the riverine corridor and hence on the extent to which ecosystem rehabilitation can take place. In areas where inner city space may not be limited but where water quality is nonetheless poor, expenditure on habitat diversity rehabilitation may be futile, assuming that water quality determines the upper limit of aquatic biodiversity.

From this perspective, then, costly expenditures on superficial improvements in aquatic habitats in highly impacted urban areas may arguably achieve better ecological ends for the same costs, if implemented in areas where there are fewer spatial constraints and, ideally, improved water quality, both of which allow for ecosystem rehabilitation on an ecologically significant scale. Again, the caveat to this statement would be the specific circumstances of each situation. Some authors have advocated the protection of important downstream systems through the creation of multiple nodes of aquatic habitat in otherwise sterile, canalized upstream areas – if implemented on a catchment level, rather than at the level of individual nodes, this may indeed be an effective strategy (Armitage et al., 2000).

Intuitively, the management of aquatic ecosystems in urban areas to meet the demands of social/recreational user groups and those of natural aquatic ecosystems would appear to be more compatible than those of management to meet engineering and ecological demands. However, beyond the level at which ecosystem failure has health and aesthetic implications, the demands of such multiple users of ecosystems are often in conflict (Day et al., 2005). The provision of aquatic habitat diversity along stream or wetland margins, for example, is often at the expense of space that could be used for walkways or the provision of broader recreational amenity (playing fields, etc.) or may simply be perceived as 'messy', by urban users who are used to open space maintenance practices that ensure neatness and straight edges over hydraulic diversity and the provision of vertical heterogeneity in cover and slope (Day et al., 2005; Day and Ractliffe, 2002). Effecting compromises in terms of allocating different nodes of a system to different uses, thus affecting the extent of area allocated to a use rather than its quality for a particular function, may be a more effective way of facilitating multiple uses of aquatic systems than compromising joint outcomes to such an extent that none of the objectives of any competing user groups are satisfactorily met.

The following section examines the role of ecological buffer areas in the protection, management and long-term sustainability of aquatic ecosystems in an urban context.

7.6 USE AND APPLICATION OF ECOLOGICAL BUFFER AREAS IN THE MANAGEMENT OF URBAN AQUATIC ECOSYSTEMS

7.6.1 Defining ecological buffers

A critical component of the management of urban aquatic ecosystems through their incorporation into corridors of open space and the delineation of these spaces on urban plans is the actual width of the planned corridors. In an urban context, surrounding land-uses are frequently incompatible with aquatic ecosystem function (Castelle et al., 1992) and may affect wetlands and river systems through increased runoff, sedimentation, introduction of chemical and thermal pollutants, diversion of water supply, the introduction of invasive and exotic species and reduced populations of indigenous wetland dependent species (Castelle et al., 1992). An effective method of reducing the impacts of development on adjacent rivers or wetlands is to provide a buffer area around the system. Wetland or riverine buffers are defined as areas that surround a wetland and abut a river bank and reduce adverse impacts to natural ecosystem functions and values from surrounding land-use (Castelle et al., 1992). Buffer areas can include both upland (terrestrial) and aquatic areas contiguous with a wetland or riparian edge. Since in many areas the latter could arguably require a buffer of their own, this section focuses on the characteristics of largely terrestrial (non-wetland or riverine) buffer areas, although it is noted that in some cases integrating the use of hardy (often artificial) wetlands with the incorporation of buffer areas provides markedly higher success rates for protecting downstream resources than either method alone (Jorgensen et al., 2000). This point should, however, be read in conjunction with comments made in previous sections regarding the need for clear prioritization of wetlands and other aquatic habitats in terms of the degree to which they lend themselves to biodiversity conservation versus those that facilitate other functions (e.g., stormwater management), without undergoing degradation or loss of biodiversity. The latter tend to be already-impacted systems (Day et al., 2005). Schulte-Hostedde et al. (2007) advise against the use of any natural wetland areas for stormwater management or buffering purposes.

7.6.2 The function of ecological buffers

Miltner et al. (2004) show in a study of streams in urban and suburbanizing landscapes in Ohio, USA, that the only sites where biological integrity is maintained, despite high levels of land-use, occur in streams where the floodplain and riparian areas are relatively undeveloped, providing buffering for natural riverine habitats. Wade et al. (1998) note the use of buffer zones as a flexible rehabilitation tool, which acts as a filter between activities in the catchment and the water body. The value of the incorporation of buffer areas at the planning level into effective catchment management strategies has been widely recognized (Large and Petts, 1994).

Even within the aquatic ecosystem itself, the previous section has shown that different users may have ecologically incompatible demands, frequently as a result of tight spatial constraints. One of the objectives of implementing effective buffer areas is to mitigate against these conflicts and to facilitate fitting the mosaic of different land-uses, impacts and ecosystems into a coherent urban layout.

From an ecological perspective, the creation or maintenance of buffer areas can facilitate any of the following functions in urban areas:

- Buffers act to mitigate against the impacts of poor water quality in both point source pollution and, importantly, diffuse runoff into rivers and wetlands, including increased levels of nutrients, sediment, organic material, faecal bacteria, heavy metals and other contaminants that occur in urban runoff. The nature of the buffer area will determine its effectiveness in addressing each of these issues. Buffers that comprise wetlands, sometimes developed artificially, may be highly effective in terms of nutrient uptake, for example. Sediment trapping may be achieved effectively by a high level of surface cover, including grass. In general, the composition of vegetation comprising the buffer area dictates the extent to which it is able to improve water quality impacts, with many vegetation communities acting as a barrier to water flow, increasing infiltration and facilitating spread of flow. As such, they encourage sediment settling and allow the sorption of nutrients and heavy metals, either through direct uptake or as a result of absorption and/or complexing within the soil and root matrix.
- Buffers can lessen the hydrological impacts often associated with hardened catchments in urban areas by improving soil infiltration capacity upstream of aquatic habitats, reducing the rate of surface discharges and thus of impacts like erosion and rapid fluctuations in water level as a result of high runoff rates. The extent to

which such functions achieve measurable levels depends however on the relative extent of the buffer area: in many urban areas, catchment hardening may be at such an extent that it overshadows the positive but localized impact of buffer areas along water courses (Houlahan and Findlay, 2004).

- Buffer areas also protect rivers and wetlands from disturbance as a result of activities taking place alongside them that may result in increased levels of noise or light, increased physical disturbance, including trampling and compaction (Castelle et al., 1992) dumping of rubble and waste and allowing the spread of seeds from alien plant material from adjacent source areas. Protection is achieved both as a function of distance from a source of impact and through the establishment of particular types of vegetation communities, which can provide visual separation between wetlands and surrounding land-use. Protection may also be rendered more effective through the installation of physical barriers to vehicle access (e.g. concrete bollards across road edges), which limit dumping and vehicle compaction, while controlled footpaths can also restrict areas of human disturbance.

- Buffers also provide space for future restoration activities along the river corridor – even where rehabilitation of highly impacted urban aquatic ecosystems is not immediately realizable. It is helpful at the planning stage to demarcate buffer areas that include sufficient space to allow realistic rehabilitation and adequate subsequent buffering of a rehabilitated system that is thus potentially more sensitive to future impacts. Failure to provide for increased buffer requirements following rehabilitation can render costly rehabilitation and remediation measures largely meaningless, if insufficient space remains to buffer them from surrounding land-use impacts.

- Buffers allow conservation of the multi-dimensional nature of the wetland or river corridor (longitudinal: upstream-downstream; vertical: river bed, interstitial environment and groundwater components; lateral: bed-bank-riparian areas; temporal: including cyclical and seasonal changes), thus ensuring the maintenance of key ecological processes (FISRWG, 2001).

- Buffers also provide habitats for species that move, during part or all of their life cycle, between aquatic, semi-aquatic and adjacent terrestrial areas: in the case of rivers, riverine corridors are often the only areas of semi-natural habitat that traverse the city and the use of adjacent space for the provision of habitat for terrestrial species allows for more effective conservation of both aquatic and terrestrial ecosystem components.

- Buffers provide longitudinal and lateral corridors for semi-aquatic fauna, and links with adjacent terrestrial areas across terrestrial or inland water ecotones, thus contributing to the concentration of biotic diversity in these ecotones and to the dispersion of animal and plant populations to adjacent terrestrial and aquatic ecosystems. Burbrink et al. (1998) note in this regard that buffer areas need to take into account the specific natural history requirements of key species for which they have been designed to protect, noting that local habitat heterogeneity and distance from core biodiversity areas are key determinants of the extent to which buffer areas actually fulfil this kind of function.

- In addition, depending on the width of the buffer area and the nature of the impacts it is intended to protect against, some buffer areas may also be able to provide additional functions, such as recreational facilities, provided that these are not in conflict with the purpose for which the buffer area has been designed.

7.6.3 Defining buffer width

Broad guidelines have been developed at both metropolitan and national levels in many countries for the imposition of buffer areas for the protection of river and wetland systems (including artificial wetlands) from urban impacts. A survey of available literature shows basic riverine buffers to range between 10 and 60 m from the river bank-full edge, depending on the quality and resilience of the receiving water body. Increasingly, however, it is also recognized that the width of the ecological buffer area should also be based on the functions it is expected to perform, such as water quality amelioration or noise protection (Castelle et al., 1992; Armstrong, 2005). These functions may vary between different rivers, and even along the same river.

The simplest application of development buffer areas around urban aquatic systems is that extended to discrete, channelized systems, often in highly developed urban areas, from which most associated floodplain wetlands have been lost. Wider buffer systems are required to protect systems that retain more of their natural hydrological functions – for example, the persistence of active floodplains. In this context it is interesting to note that in fact, the floodplain system itself may warrant allocation of a buffer area around it, rather than presenting the floodplain as part of the buffer area of the riverine system. This point highlights the importance of linking riverine systems with their associated wetlands in spatial planning in order to ensure protection of the most sensitive component, or that closest to the source of impact, by imposing buffer areas.

An array of published and unpublished literature relates to the setting of ecological buffers around wetlands. Although most of the literature is based on best-guess estimates, in some cases, scientific research has produced data to establish specific criteria in terms of the nature and extent of ecological buffers intended to perform specific functions. Semlitsch and Bodie (2003) showed that core wetlands need to be edged by terrestrial habitats of up to 290 m in width if their purpose is to sustain key amphibian and reptile life-history functions. As part of this study, these authors demonstrated that the 15 to 30 m wide buffer areas often utilized around wetlands are inadequate to protect these classes of wetland-associated animals from edge impacts and habitat loss. In fact, in such cases the required buffer area clearly functions as habitat in its own right, as well as buffering wetland areas from specific impacts. In the case of many wetlands, particularly urban ones the provision of this level of habitat may be irrelevant, as the wetlands themselves no longer support these levels of fauna.

Castelle et al. (1992) identify the following criteria in determining buffer size:

• *Resource functional value (i.e., the sensitivity of the wetland or river resource to specified impacts)*: more sensitive streams or wetlands, or those that are highly threatened, generally require wider buffers than those that have lower sensitivity, although it is cautioned that 'sensitivity' may not be a generic characteristic of an aquatic ecosystem, which may be highly sensitive to certain impacts, and less sensitive to others. Thus sensitivity ratings for particular identified impacts need to guide an assessment of this criterion. The ability of an aquatic ecosystem to process impacts internally should also be assessed – dense *Typha capensis* reedbeds, for example, are often highly resilient in terms of their sensitivity to impacts like sediments, nutrient enrichment and other water quality impacts (Hall, 1990), while Ziegler et al. (2006) provide one example of research showing the efficacy of hardy Cyperaceae sedges (in this case *Fimbristylis aphylla*) in terms of sediment trapping and nutrient uptake. Such wetlands would not necessarily require a wide ecological buffer to protect them from these particular impacts, whereas a fast-flowing mountain stream or a more sensitive wetland type might be highly sensitive to these particular categories of impact.

- *Intensity of adjacent land-uses (i.e., the nature of anticipated impacts)*: buffer widths should respond to the kind of impact anticipated or present in a system. Thus in the case of a low-density development, where attention has been paid to landscaping with indigenous natural vegetation, where the likelihood of nutrient-enriched runoff into the system is low, as a result of internal policies that limit the application of fertilizers and the over-irrigation of planted areas, recommended buffer areas should be substantially less than in cases where similar types of receiving water bodies are impacted by high-intensity developments associated with a large area of hardened surface relative to total size and a high likelihood of generating polluted runoff (e.g. from car parks, irrigated and fertilised planted areas or lawns, or direct discharge of contaminated stormwater).

- *Buffer characteristics (i.e., the efficacy of the buffer in ameliorating impacts)*: the physical, chemical, hydrological and botanical characteristics of the buffer area are crucial indicators of buffer efficacy in mitigating against identified impacts. Factors such as soil type and porosity, saturation pattern and organic content, slope, rainfall intensity, groundwater influences (and interactions between surface and groundwater) and vegetation (structure and function, including aspects such as height, cover, root system and species-specific properties in terms of uptake of key nutrients or other pollutants) are all vital considerations in determining buffer efficacy. Kadlec and Knight (1996) summarize some of the properties of different wetland soil types that render them more or less efficient in mitigating against different kinds of runoff. The organic content of soil plays a key role in determining its impact on water quality, with highly organic soils often being effective in water retention, while humic acids increase the potential for forming biologically unavailable complexes with certain heavy metals. Aerobic soils tend to retain phosphorus as insoluble complexes within a wetland. By contrast, under anaerobic conditions, which are often associated with saturation, phosphorus becomes biologically available.

- *Required buffer functions (the purpose the buffer is intended to fulfil)*: depending on the nature of surrounding land-use, buffer widths might be intended to perform a range of functions or to provide effective protection against a single identified impact. As a general rule, the more impacted the receiving water body is, including its natural riparian fringe, the more likely it is that a buffer will need to address single issues, which seldom relate to maintenance of habitat or prevention of disturbance, but rather generally centre on issues like water quality amelioration or dissipation of surface runoff. In less impacted systems, where the system has at least remnant ecosystem function, providing habitat to communities of plants and animals that move along or within the riverine or wetland corridor, the functions of a buffer area may need to extend to multiple uses.

- *Ownership or control of the buffer area*: Castelle et al. (1992) note the positive impact on buffer function of their location on land owned or managed by individuals with a clear understanding of the rationale for their existence, commenting also in their review that buffer areas located within residential areas even tended to result in a loss of natural vegetation communities and the development of lawns over time.

Based on the above criteria, buffer widths between 10 m – sediment removal, erosion control and reduction of dissolved nutrients (Naiman and Decamps 1990) – and 106 m (species diversity) were recommended for the protection of wetlands, although Castelle et al. (1992) caution that in practice, buffers of less than 20 m in width are generally inadequate for the protection of wetlands.

7.6.4 The moderating effect of urban layout and land-use on requirements for buffers

Subtleties in urban design can also play a role in determining the efficacy of a buffer in its capacity for protection of an aquatic ecosystem. For example, urban layouts that allow for the edging of a wetland and/or riparian corridor by a road are often more effective than a similar buffer width that is incorporated within, or edged by, the boundary of adjacent properties. This is because the hard edge of a road provides a clearly discernible management edge and may be less subject to edge effects such as the expansion of artificial cultivated features (lawns, etc.) that extend into the buffer over time (Castelle et al., 1992). The buffer area, thus separated from private ownership, is also easier to police and control. Inclusion of such aspects of urban layout in a development is one means whereby the ecological requirements for a buffer area become negotiable by improving the quality of the buffer area created, and thus allowing reduction in its extent.

Since the primary purpose of the ecological buffers described above is to protect the aquatic ecosystem from surrounding impacts, particular attention should be given to the use of riverine corridors for additional purposes, where these may be in conflict with the primary function of the buffer area. Human disturbance in the form of noise and trampling might be indirect negative consequences associated with recreational uses of buffer areas (Emmons and Oliver Resource Incorporated, 2001), with other potential effects on buffer efficacy if recreational use also dictates plant growth forms – grass versus multi-layered shrubs, reeds or wooded areas (Day et al., 2005).

Thus tradeoffs in spatial allocations to buffer areas are also often necessary: where multiple uses of buffer areas will take place, either wider buffer areas may be necessary, or a loss in buffer efficacy with concomitant effects on aquatic ecosystem integrity would be anticipated. Alternatively, attention can be paid to the active rehabilitation of aquatic habitats and improving the efficiency of sections of the buffer that are adjacent to the aquatic system itself, in order to accommodate a loss of function elsewhere. This concept can be incorporated into the application of another concept of development planning: namely, the notion of biodiversity offsets. Biodiversity offsets are conservation activities intended to compensate for the residual, unavoidable harm to biodiversity caused by development projects (IUCN, 2004). In the case of offsetting residual impacts to aquatic habitats in urban areas, efforts to improve the efficiency of buffer functions, such as water quality amelioration and sediment trapping at key nodes along a river system or within a catchment, where sufficient space is available for such measures to be effective, might be incorporated into long-term spatial and development planning.

7.6.5 Realities of implementing requirements for buffer areas

While the use of ecological buffers is widely advocated in the protection of urban aquatic ecosystems (Castelle et al., 1992; Large and Petts, 1994; Wade et al., 1998;

Armstrong, 2005), the spatial implications of buffer allocation should not be underestimated, and clear delineation of buffer areas on urban land-use plans should be encouraged in order to ensure that expectations of development opportunities are realistic, if they are to take place within a framework of sustainable aquatic ecosystem protection and/or management. Schulte-Hostedde et al. (2007) note that while varied buffer widths may allow for a more effective protection of aquatic habitats, a fixed buffer width for all systems is often perceived as fairer, easier to administer and more easily understood by members of the public.

Implicit in long-term planning for the incorporation of buffer areas into the urban framework is the need for an accurate delineation and inventory of existing systems, their ecological (and other) importance, and their sensitivity to different identified impacts. It is only through a clear understanding of the processes and patterns that exist in or are dependent on particular urban aquatic ecosystems that provision can be made for their incorporation into and protection from the urban environment. Even in this context, it is noted that incorporation into a framework does not necessarily ensure protection from future impacts. Thus policies that advocate the use of buffers, need to be backed up by effective and implementable legislation affording them legal protection into the future.

7.7 TOOLS TO ASSIST IN THE INTEGRATION OF, AND TRADEOFFS BETWEEN, SOCIAL, ECOLOGICAL, ENGINEERING AND ECONOMIC REQUIREMENTS IN URBAN AREAS

This chapter has focused on the need for an integrated approach to the resolution and management of ecological, social, economic and engineering concerns in urban areas, and highlighted the importance of a holistic approach to spatial planning in achieving this objective and reducing areas of conflict between different user groups. Ongoing monitoring of specified criteria, both in natural ecosystems and in the built environment, including its ecological, social and economic sectors, is an important component of this approach. However, fundamental to the interpretation of measured criteria is the need for indices to assist in formulating decisions that adequately take into account the (often conflicting) concerns of different sectors, and allow integration of ecological, hydrological and geohydrological data with spatial, social and other issues.

Monitoring aquatic habitats in urban areas is also necessary in order to inform future management and intervention decisions relating to the affected system, as well as to other systems for which intervention or management is planned in the future. Ideally, the kinds of indices that are developed or applied should include predictive capacities in order to allow the evaluation of possible synergistic and/or antagonistic effects on other systems or communities, as well as allowing comparison between different systems and, over time, to allow ongoing fine-tuning of intervention measures, the impact of which can be assessed against measured values.

7.8 CONCLUSIONS

This chapter has discussed in some detail the critical need for issues relating to the maintenance, conservation or rehabilitation of aquatic habitats in urban areas to be incorporated into urban planning, so that allocations of open space can take into consideration

the need for ecologically meaningful corridors across the urban area and between remnant nodes of natural habitats within the urban matrix. In addition, the need to implement protective measures on a city-wide basis aimed at minimizing impacts on aquatic
habitats as a result of adjoining land-uses means that the allocation of sufficient area for
the development of buffer areas between the edge of aquatic habitats and adjacent urban
land-uses should also be prioritized, if medium-term objectives of improving levels of
biodiversity protection and ecosystem status are to be achieved realistically. Without the
integration of ecological, social and economic needs at these levels, efforts to safeguard
biodiversity in urban areas are unlikely to be sustainable. At the same time, monitoring
of the efficacy of the measures implemented is also important within a framework that
allows for flexibility in the long-term approach, so that improvements in methodologies
may be made over time, and measures that do not meet their objectives may be revised
in a timely fashion without permanent ecological repercussions.

The use of multi-disciplinary teams in long- and short-term development planning
and management decisions has been highlighted as essential if integration between
quite different spheres of the urban landscape is to be achieved. Moreover, cross-
institutional collaboration (e.g., between universities, municipalities, consultants and
planning departments) is also essential for ensuring adequate communication of
issues, potential intervention strategies and practical limitations in implementation.

Designing cities compatible with their ecological surroundings will have considerable impacts on the establishment of sustainable landscapes. This is because sustainable landscapes are responsive to human needs, fostering human understanding of
natural and historical processes associated with a place and the protection of biodiversity and global sustainability (Hahn, 2002; Ahern and Boughton, 1994). It is thus
essential that urban planning practices have to be integrated with ecosystem functioning and human interventions in a multi-dimensional perspective (Tezer, 2005).

The increase in population and its concentration in urban areas is predicted to continue
with greater rates and more severe impacts this century, especially in developing countries,
and mega cities of the developing world are likely to illustrate this trend in their hinter-
lands too (UN, 2004). Biodiversity is highly vulnerable to threats from such uncontrolled
development and the unsustainable consumption habits of contemporary life styles.
Global policies promote different initiatives for the common goal of 'global sustainability
with local commitment' (UNCTAD, 1992). The emergence of an understanding of the
extent of environmental degradation and its multiple life-threatening impacts means that
compliance of different countries with their global commitments should be mandatory.

Although many countries are signatories of important global treaties, including the
RAMSAR convention and others not listed here, international agreements on their
own are insufficient for fostering biodiversity conservation. Their efficacy depends on
the extent to which they are incorporated into development plans prepared by local
and/or central government authorities, and secured by participatory development programmes at local and global levels (Tezer, 2005).

REFERENCES

Ahern, J. and Boughton, J. 1994. Wildflower Meadows as Sustainable Landscapes. R.H. Platt,
    R.A. Rowntree and P.C. Muick (eds) *The Ecological City*. The University of Massachusetts
    Press, Amherst, pp. 172–87

Alfsen-Norodom, C., Boehme, S.E., Clemants, S., Corry, M., Imbruce, V., Lane, B.D., Miller, R.B., Padoch, C., Panero, M., Peters, C.M., Rosenzweig, C., Solecki, W. and Walsh, D. 2004: Managing the megacity for global sustainability: The New York Metropolitan Region as an urban biosphere reserve. Ann. *New York Acad. Sci.*, 1023, pp. 125–41, doi:10.1196/annals.1319.005.

Armitage, P.D and Cannan, C.E. 2000. Annual changes in summer patterns of mesohabitat distribution and associated macroinvertebrate assemblages. *Hydrological Processes* 14 (16–17): pp. 3161–79 Nov-Dec 2000

Armstrong, A.J. 2005. Guidelines to address biodiversity concerns in industrial and commercial timber plantations in kwaZulu-Natal. Biodiversity Division. Ezemvelo KZN Wildlife.

Beatley, T. and Manning, K. 1997. *The Ecology of Place, Planning for Environment, Economy and Community*, Island Press, Washington DC.

Bostelmann, R., Braukmann, U., Briem, E., Fleischhacker, T., Humborg, G., Nadolny, K.S. and Weibel, U. 1998. An approach to Classification of natural streams and floodplains in southwest Germany. L. De Waal, A. Large and P. Wade (eds) *Rehabilitation of Rivers*, Chichester, John Wiley & Sons, 1998, pp 31–55.

Braioni, A. and Braioni, M.G. 2002. Fluvial Landscape evaluation: a method of analysis suitable to ecocompatible planning. G. Ceccu, U. Maione, B. Lehto, R. Monti, A. Paoletti, M. Paoletti, U. Sanfilippo. (eds) 2nd Conference "New Trends in Water and Environmental Engineering for Safety and Life. Eco-compatible Solutions for Aquatic Environments (Capri 24–27 June 2002). Centro Studi Deflussi Urbani. D.I.I.A.R Politecnico di Milano. ISBN 88-900282-2-X: 1-14.

Braioni, MG., Braioni A. and Salmoiraghi G. 2006a. Model and tools for the integrated management and planiing of the river – fluvial corridor. *Ecohydrology & Hydrobiology*, 5 (4) pp. 323–36.

Braioni, M.G., Braioni, A. and Salmoiraghi, G., 2006b. A model for the integrated management of river ecosystems. *Verh. Internat.Verein. Limnol.*, 29 (4): pp. 2115–23.

Burbrink, F., Phillips, C. and Heske, E. 1998. A riparian zone in southern Illinois as a potential dispersal corridor for reptiles and amphibians. *Biological Conservation.* 86: pp. 107–15.

Castelle, A.J., Johnson, A.W. and Conolly, C. 1992. Wetland and Stream Buffer Size Requirements – A Review. *Journal of Environmental Quality* 23: pp. 878–82.

CBD. 1992. Convention on Biological Diversity. United Nations Environment Programme (http://www.biodiv.org/doc/legal/cbd-en.pdf).

Conway, T.M. and Lathrop, R.G. 2005. Alternative land use regulations and environmental impacts: assessing future land use in an urbanizing watershed. *Landscape and Urbana Planning* 71 (1): 1–15 FEB 28 2005.

Council Directive 85/337/EEC of 27 June 1985 on the assessment of the effects of certain public and private projects on the environment.

Council Directive 92/43/EEC of 21 May 1992 on the conservation of natural habitats and of wild fauna and flora.

Day, E. and Ractliffe, G. 2002. *Assessment of river and wetland engineering and rehabilitation activities within the City of Cape Town. Realisation of project goals and their ecological implications. Volume 1: Assessment process and major outcomes.* Report to Catchment Management. Cape Metropolitan Council. City of Cape Town.

Day, E., Ractliffe, G. and Wood, J. 2005. An audit of the ecological implications of remediation, management and conservation or urban aquatic habitats in Cape Town, South Africa, with reference to their social and ecological contexts. *Ecohydrology and hydrobiology.* Vol 5:(4), pp. 297–309.

Dini, J.A. and Cowan, G.I. 2000. Classification system for the South African wetland inventory. Second Draft prepared for the South African Wetlands Conservation Programme, DEAT.

Directive 2000/60/EC of the European Parliament and of the Council of 23 October 2000 establishing a framework for Community action in the field of water policy.

DiSano, J. 1995. Indicators of Sustainable Development: Guidelines and Methodologies, UN Division for Sustainable Development, http://www.un.org/esa/sustdev/natlinfo/indicators/isd.htm, Accessed: 11 August 2005.

EEA. 2003. *An Inventory of Biodiversity Indicators in Europe 2002*, Technical Report No.92, Prepared by Ben Delbaere, Project Manager Ulla Pinborg, European Environment Agency, Copenhagen.

Emmons and Oliver Resources Incorporated. 2001. Benefits of wetland buffers: A study of functions, values and size. Prepared for the Minnehaha Creek Watershed District. Minnesota.

Finlayson, C.M. and van der Falk, A.G. 1995. Wetland classification and inventory: a summary. *Vegetatio*. 118, pp. 185–92.

FISRWG. 2001. Stream Corridor Restoration. Principles, Processes, and Practices. ISBN-0-934213-59-3. http://www.nrcs.usda.gov/technical/stream_restoration.

Gopal, B.B. and Wetzel, R.G. (eds). 1995. Limnology in Development Countries. SIL International Association of theoretical and applied limnology, vol. 1.

Gopal, B.B. and Wetzel, R.G. (eds). 1999. SIL International Association of theoretical and applied limnology, Vol. 2.

Gopal, B.B. and Wetzel, R.G. (eds). 2001. Limnology in Development Countries. *SIL International Association of theoretical and applied limnology*, Vol. 3.

Gopal, B.B. and Wetzel, R.G. (eds). 2004. Limnology in Development Countries. *SIL International Association of theoretical and applied limnology* Vol. 4.

Gyllin, M. 1999. Integrating Biodiversity in Urban Planning, *Proceedings of the Gothenburg Conference: Communication in Urban Planning*, Gothenburg, Sweden. http://www.arbeer. demon.co.uk/MAPweb/Goteb/got-mats.htm, Accessed: 26 April 2006.

Hahn, E. 2002. Towards Ecological Urban Restructuring: A Challenging New Eco-cultural Approach, *Ekistics*, 69, (412–14), pp. 103–15.

Hall, D. 1990. The biology and control of *Typha capensis* in the south-western Cape Flats. Freshwater Research Unit Report to the City of Cape Town.

Houlahan, J.E. and Findlay, C.S. 2004. Estimating the 'critical' distance at which adjacent land-use degrades wetland water and sediment quality, *Landscape Ecology* 19 (6): pp. 677–90 Aug 2004.

IUCN. 2004. The IUCN Species Survival Commission. IUCN Red List of Threatened. http://www.iucn.org/themes/ssc/red_list_2004/GSAexecsumm_EN.htm.

Jasperse, P. 1998 Policy Networks and the Success of Lowland Stream Rehabilitation Projects in the Netherlands. L. De Waal, A. Large and P. Wade (eds) *Rehabilitation of Rivers*, Chichester: John Wiley & Sons, pp 14–29

Jorgensen, E., Canfield, T.K.J. and Kutz, F. 2000. Restored riparian buffers as tools for ecosystem restoration in the Maia; processes, endpoints and measures of success for water soil flora and fauna. *Environmental monitoring and assessment*. 63: 199–210. Kluwer Academic Publishers. Netherlands.

Kadlec, R. and Knight, R. 1996. Treatment wetlands. Lewis Publishers. New York, pp 893.

Large, A.R.G. and Petts, G.E. 1994. Rehabilitation of river margins. P. Calow and G.E. Petts (eds). *The Rivers Handbook. Volume 2*. Blackwell Scientific Publications. Oxford, pp 401–18.

Laros, M. 2004. *Biodiversity prioritization project*. Report prepared for the City of Cape Town in association with the Biodiversity and Conservation Planning Department, University of the Western Cape, and the Freshwater Consulting Group.

Lee, D.J. and Dinar, A. 1995. Review of Integrated Approaches to River Basin Planning, Development and Management. *Policy Research Working Paper 1446*, World Bank.

Levy, J.M. 2000. *Contemporary Urban Planning*, Fifth Edition, Prentice Hall.

Matsuura, K. 2005. Message from Mr. Koïchiro Matsuura, Director-General of UNESCO on the occasion of World Environment Day, 5 June 2005, http://www.unesco.org/mab/ecosyst/urban/doc.shtml, Accessed: 25 July 2005.

McHarg, I. 1969. *Design with Nature*, Garden City, NY, Anchor.

Miltner, R.J., White, D. and Yoder, C. 2004. The biotic integrity of streams in urban and suburbanizing landscapes. Landscape and urban planning. 69: pp. 87–100.

Naiman R.J. and Decamps, H. (eds). 1990. *The Ecology and management of aquatic – terrestrial ecotones*. Man and Biosphere Series, 4.

OECD. 2002. OECD *Territorial Reviews*, Canada.

Pauly, D. and Christensen, V. 1995. Primary production required to sustain global fisheries. *Nature* 374 (6519): 255–57 Mar 16.

Peck, S. 1998. *Planning for Biodiversity: Issues and Examples*. Washington DC: Island Press.

Platt, R.H. (ed.) 2006. *The humane metropolis: People and nature in the 21ˢᵗ century city*. UMass Press, Amherst & Boston, Lincoln Institute of Land Policy, Cambridge.

Portney, K.E. 2003. *Taking Sustainable Cities Seriously. Economic Development, the Environment and Quality of Life in American Cities*, MIT Press, Cambridge, MA.

Pringle, C. 2003. The need for a more predictive understanding of hydrologic connectivity. Aquatic Conservation-Marine and Freshwater Ecosystems 13 (6): 467–71, Nov/Dec, 2003.

Roux, D.J, Nel, J., McKay, H. and Ashton, P. 2006. Discussion paper on cross-sector policy objectives fro conserving South Africa's inland water biodiversity. Water Research Commission Report No TT 276/06. Pretoria. South Africa.

Schulte-Hostedde, B., Walters, D., Powell, C. and Shrubsole, D. 2007. Wetland management: An analysis of past practice and recent policy changes in Ontario. *Journal of Environmental Management*. 82: (1) pp. 83–94.

Semlitsch, R.D. and Bodie, J.R. (2003). Biological Criteria for Buffer Zones around Wetlands and Riparian Habitats for Amphibians and Reptiles. *Conservation Biology*, Vol. 17, No. 5. pp. 1219–28.

Sukopp, H. and Weiler, S. 1988. Biotop Mapping and Nature Conservation Strategies in Urban Areas of the Federal Republic of Germany, *Landscape and Urban Planning*, 15, pp. 39–58.

Tezer, A. 2005. The Urban Biosphere Reserve (UBR) concept for sustainable use and protection of urban aquatic habitats: case of the Omerli Watershed, Istanbul, *Ecohydrology & Hydrobiology*, Vol. 5, No. 4, pp. 311–22.

UN. 2004. World Population Prospects. The 2004 Revision. United Nations. Population Division. http://esa.un.org/unpp/.

UN-CSD. 2005. Convention on Sustainable Development, Theme Indicator Framework, http://www.un.org/esa/sustdev/natlinfo/indicators/isd.htm Accessed: 24 July 2005.

UNCTAD. 1992. *World Investment Report 1992: Transnational Corporations as Engines of Growth*, United Nations Publication, New York and Geneva.

Wade, P.M., Large, A.R.G. and de Waal, L.C. 1998. Rehabilitation of degraded river habitat: An introduction. L. De Waal, A. Large and P. Wade (eds) *Rehabilitation of Rivers*, Chichester: John Wiley & Sons, pp 1–10.

Ward, JV, Tockner, K and Schiemer, F. 1999. Biodiversity of floodplain river ecosystems: Ecotones and connectivity. *Regulated Rivers-Research & Management* 15 (1-3): pp. 125–39, Jan-Jun, 1999.

Zalewski, M. 2000. Ecohydrology – the scientific background to use ecosystem properties as management tools towards sustainability of water resources (guest editorial). *Ecological Engineering*, Vol. 16, pp. 1–8.

Zerbe, S., Maurer, U., Schmitz, S. and Sukopp, H. 2003. Biodiversity in Berlin and its Potential for Nature Conservation, *Landscape and Urban Planning*, 62, pp. 139–48.

Ziegler, A.D., Negishi, J. and Sidle, R.C. 2006. Reduction of stream sediment concentration by a Riparian buffer: Filtering of road runoff in disturbed headwater basins of Montane mainland Southeast Asia. *Journal of Environmental Quality*. 35 (1): 151–62 Jan-Feb 2006.

Chapter 8

Human health and safety related to urban aquatic habitats

Tuula TUHKANEN

Tampere University of Technology, Institute of Environmental Engineering and Biotechnology, P.O. Box 541, FI-33101 Tampere, Finland

8.1 INTRODUCTION

Water bodies have always been an essential element for human societies by providing drinking water, food, transportation, and recreational areas and by playing important cultural roles. However with intensifying urbanization, they have also become sinks for liquid and solid wastes. The modification of urban aquatic habitats by various stressors usually results in their pollution, physical simplification, degradation of the biological structure and ecosystem functioning. Where such degradation exceeds ecosystem capacity, it not only impairs ecosystem services but also, potentially, public health. However, in the overall assessment, properly protected and managed urban aquatic habitats can be valuable assets that greatly enhance the quality of life in cities.

Among the most important services provided by aquatic ecosystems are those of importance for humans functioning in cites, such as water and air purification, drought and flood mitigation, waste detoxification and decomposition, biodiversity maintenance and climate stabilization. They also provide aesthetic amenities and intellectual stimulation that lift the human spirit and improve health. Causal links between the environmental change and human health are complex, because they are often indirect, displaced in time and space, and depend on a number of modifying forces (UN, 2005). However, the link between environmental quality and human health is widely recognized and considered to be very important.

Rapid urbanization produces pollution loads, which may degrade aquatic ecosystem functions and impair water quality. As population densities increase, the health-related characteristics of waters, such as the presence of disease-causing agents, pathogenic micro-organisms and anthropogenic chemicals, become more significant for the public health. Most serious environmental and health impacts resulting from water pollution can be kept under control by the technical solutions discussed in Chapter 4, combined with measures increasing aquatic ecosystems' resilience and absorbing capacity (see Chapters 2 and 6). However, many challenges still remain, even the basic ones related to safe drinking water supply and improved sanitation, especially in developing countries. Thus, the main issues addressed in this chapter deal with the management of urban aquatic habitats with respect to providing required goods (water) and water services to the urban population with the overall goal of improving human health and the quality of life.

8.2 WATER BODIES AS SOURCES OF RAW WATER AND WASTEWATER SINKS

Modern centralized waste collection and treatment systems have brought about significant improvements in public health in developed countries, which can afford such solutions. These systems, however, convey urine, faeces and other wastes flushed into, and transported in, the sewage system by huge amounts of high quality water, still seriously diminishing its resources. It has been estimated that just 20% of all wastewater produced in the world is treated, and even in the European Union (EU) the coverage of proper treatment is just 50% (World Water Assessment Programme, 2006). Polluted water affects downstream activities, the withdrawal of raw water by municipalities, bathing and other recreation uses of water bodies (Tambo, 2002). The pollution load also adversely affects fish populations and the functioning of aquatic ecosystems. Even treated effluents from conventional wastewater treatment plants may contain chemicals of concern as well as micro-organisms. Depending on the treatment processes and removal efficiencies, some residuals always pass through the treatment process and end up in the terrestrial or aquatic environments. Currently, for drinking water production, many communities in Europe and worldwide use water sources that may contain a significant portion of treated wastewater, or may be affected by diffuse pollution from agriculture or low-density housing developments. In many parts of the world, these effluents comprise small (10 to 25%) to significant (>75%) portions of the natural flow that is used downstream for potable water abstraction (Bixio et al., 2006). During seasonal or prolonged droughts, the contribution of wastewater effluents to raw sources of drinking water further increases.

The pollution of surface waters makes groundwater resources a significant raw water source. It can also contain wastes due to intentional or unintentional infiltration into the soil. Landfills are a major threat to groundwater quality, because particularly older and poorly designed and constructed landfills produce leachates that can infiltrate into soils and be transported into water bodies. The case study of the Phoenix Metropolitan Area in Arizona, USA (see Chapter 9) shows the importance of deep groundwater sub-basins in supplementing surface water supply, which has been taking place since the early 1900s. Declining water tables and pollution occurring in some places over the last several decades has raised many concerns and forced the implementation of special management strategies.

Widely used decentralized sanitation systems, which are based on the use of water toilets in rural or urban areas, cause numerous local problems of groundwater contamination. In typical decentralized systems, wastewater is collected in subsurface septic tanks where the suspended matter settles on the bottom and the effluent is discharged into the environment, sometimes poorly treated. Septic tank overflow can contaminate nearby domestic water wells (Szabo, 2004). The intentional infiltration of wastewater from septic tanks into soils is a common practice in rural areas of developed countries, such as Sweden and Finland, even though the treatment of septic tank effluents became mandatory in Finland in 2004. If the infiltration bed of the septic tank is properly located, constructed and maintained, most of the organic matter and pathogens are adsorbed, degraded or inactivated in the soil before the pollution plume reaches the groundwater extraction point. The wastewater-induced clogging of the soil can increase the soil biogeochemical activity and enhance sorption, biotransformation of organic constituents and die-off and inactivation of bacteria and viruses (van Cuyk el al., 2001). When soil capacity has been exceeded, harmful components can enter and contaminate groundwater (Jacks et al., 2000). Microbiological and chemical contamination of drinking water wells may also occur due to unintended wastewater disposal, accidental leaks (e.g., exfiltration from leaky sewers) and leaching of human excreta and animal manure (Scandura and Sobsey, 1997; Szabo, 2004).

8.3 MICROBIOLOGICAL RISKS AND THEIR CONTROL

8.3.1 Drinking water

The risk of waterborne infections diseases depends on the number of pathogens and their dispersion in water, the infective dose required, and the susceptibility of the exposed population (Schönning and Stenström, 2004). The most common human microbial pathogens in water are enteric in origin, and their number is proportional to the health of the local human community. The pathogenic micro-organisms enter the environment in the faeces of infected hosts and can get into the drinking water either through defecation or by water contamination with sewage or runoff from contaminated soils or surfaces. The enteric pathogens in water include viruses, bacteria, protozoa and helminths.

More than 120 different types of viruses can be excreted in faeces; the most common of which are enteroviruses, rotaviruses, enteric adenovirus and the human calicivirus group (noroviruses). Hepatitis A is also a pathogenic virus of great concern, which is related to water- and food-borne outbreaks. Among bacteria, *Salmonella*, *Campylobacter* and Enterohaemorrhagic *E.coli* (EHEC) are particularly important in developed countries. In areas with insufficient sanitation and sewage treatment, typhoid fever (*Salmonella typhi*), cholera (*Vibrio cholerae*) and *Shigella* are also common. The parasitic protozoa, *Cryptosporidium parvum* and *Giardia lamblia*, which originate very often from domestic livestock and cattle in particular, have caused some significant disease outbreaks in developed countries (Hunter and Thompson, 2005). They are highly resistant to disinfection and have to be controlled preferably by prevention of surface contamination by runoff from farms and pastures. Risk control is based on the multi-barrier principle: maintaining good quality raw water source and treating the raw water by processes, which should include optimized chemical coagulation, clarification, sand filtration and disinfection (Betancourt and Rose, 2004). In developing countries, helminth infections are very common. Millions of people suffer from *Ascaris*, Hookworm, *Schistosoma haematobium*, *S. japonicum* and *S. mansoni*. Urine, on the contrary, is practically free of harmful micro-organisms before becoming contaminated by faecal matter. The pathogens known to be excreted in urine are *Leptospira interrogans*, some forms of Salmonella and *Schistosoma haematobium*.

Examples of the most common causes of death, many of which are directly or indirectly related to water, are listed in Table 8.1. Furthermore, enteric diseases increase vulnerability to other lethal conditions, such as malnutrition, tuberculosis, HIV/AIDS and respiratory infections, and other diseases, such as malaria, are water-based (carried by mosquitoes breeding in stagnant water). The disease burden can be measured by the time lost by Deaths and Disability-Adjusted Life Years (DALYSs).

In countries with low income and insufficient or no drinking water treatment, the risk of pathogens in drinking water is continual. In developed countries local outbreaks of bacterial and viral gastroenteritis, Giardiasis and Cryptosporidiosis occur occasionally, usually after a sudden change of quality of raw water or a failure of the treatment process.

The survival of pathogens outside of the host depends on many factors. In untreated water, sunlight, thermal effects, sedimentation, filtration and predation reduce microbial levels. *E. coli* is used as a microbial indicator of the occurrence of faecal contamination. It is of great importance that indicator organisms survive in the environment and in treatment processes in the same way as the pathogenic organisms. It is known,

Table 8.1 Global burden of diseases: in 2002

	Total number of deaths		DALYs
	%	millions	%
Infectious and parasitic diseases	19.1	10.9	23.5
Respiratory infections	6.9	3.9	6.3
HIV/AIDS	4.9	2.8	5.7
Diarrhoeal diseases	3.2	1.8	4.2
Tuberculosis	2.7	1.7	2.3
Malaria	2.2	1.3	3.1

Source: World Health Report. 2004b.

however, that *E. coli* may die off more quickly than other enteric bacteria, viruses and micro-organisms, cysts and oocysts.

The municipal sewage treatment system has the advantage that it collects hazardous and highly contaminated wastewater at a central treatment plant. Most of the pathogenic micro-organisms are removed by adsorption onto sludge, or they become inactivated by bacterial competition. The water normally requires effective treatment before it is considered suitable for reuse or for discharge into the environment, since some of the viruses and parasites are highly resistant to the conditions of the biological treatment. Conventional activated sludge treatment removes up to 90% of the pathogenic organisms, and in most of the European countries, there is no further disinfection of wastewaters. On the contrary, in the US, chlorination (sometimes followed up by dechlorination) is widely practised prior to the discharge of the treated effluent into receiving water bodies. On the other hand, tertiary treatment with possible additional filtration and disinfection stages increases the removal of pathogens by several orders of magnitude and more effectively. The formation of disinfection by-products in chlorination is, however, a significant problem, and consequently, alternative disinfection methods using UV irradiation or ozonation are used.

When wastewater is discharged into surface water bodies, solar irradiation may play an important role in the elimination of micro-organisms and organic pollutants, in addition to microbial degradation and sorption (Toze, 2006). Several studies have demonstrated the significant role of constructed wetlands, applied as tertiary systems for wastewater treatment (effluent polishing), in the removal of enteric pathogens and the reduction of their survival (Karim et al., 2004; Williams et al., 1995). They are efficient at removing faecal colifoms, bacteria (e.g., *Escherichia coli*, Salmonella) and viruses. Their efficiency is strongly modified by temperature, overflow and the presence of vegetation. In the northern hemisphere, biodegradation and photodegradation processes are significantly hindered during the cold and dark seasons with ice and snow cover (Vieno et al., 2005).

In addition to drinking water pathways, there are other routes of exposure. Using water containing harmful chemicals or micro-organisms for irrigation also increases the risk of adverse health effects. Harmful substances or organisms may endanger several contact groups: farmers and their families, residents living nearby, and handlers, sellers and consumers of the agricultural products. Disease transmission may occur not only through contact or the ingestion of contaminated water or produce, but also through bioaerosols entering the human body through the mouth and lungs (Carducci et al., 2000).

Bioaerosols are small airborne particles that penetrate into lungs. Field studies indicate that wastewater treatment plants, compost and the spreading of sludge and wastewater onto fields can create a cloud of bacterial and viral bioareosols. Identified strains include, for example, *Escherichia coli*, *Clostridium perfringens*, *Staphylococcus aureus* and *Enterococcus* sp. The concentrations of bioaerosols have been measured at more than 400,000 colony forming units per cubic metre (CFU/m^3). All activities involving mechanical aeration, nebulization and sewage handling are potential risks (Fraccchia et al., 2006). Workers and residents living in the vicinity of these locations are exposed to particularly high bacterial concentrations (Sahu et al., 2005).

8.3.2 Bathing waters

The general public and policy-makers are concerned about the discharge of untreated or poorly treated municipal sewage into bathing waters (WHO, 2003). Sewage contains micro-organisms and pathogens of several human diseases, which can be transferred to the exposed population and cause, most commonly, gastrointestinal and upper respiratory infections. In developed countries, the level of micro-organisms can be reduced by advanced wastewater treatment, such as filtration and post-disinfection. Under such circumstances, the role of diffuse bacterial pollution originating in upstream catchments is becoming increasingly important. The implementation of the new EU Bathing Water Directive (EC, 2006), which will become effective no later than at the beginning of 2008, specifies two indicator organisms to be assessed: intestinal enterococci and *Escherichia coli*. Following the assessment, bathing waters will be classified as poor, sufficient, good and excellent, with sufficient representing the minimum acceptable level. The Directive, however, has been criticized as being too costly compared to the public health and sociological benefits which it can bring about (Geourgiou and Bateman, 2005).

In some regions, dangerous aquatic organisms are encountered during visits to freshwater or coastal ecosystems and can cause health problems, injury or death. These include both non-venomous organisms like sharks and alligators and venomous vertebrates and invertebrates like jellyfish, sea anemones and sea snakes (WHO, 2003).

8.3.3 Disease vectors

Many disease vectors are typically small and otherwise harmless animals, such as mosquitoes. Female mosquitoes require human or animal blood to develop their eggs. In the process of taking blood, mosquitoes can ingest pathogens causing malaria from an infected person or animal and pass the pathogen onto the next person. All mosquitoes go through an aquatic larval stage, generally using standing or slowly running clean freshwater for breeding sites. The destruction of habitats and breeding sites reduces adult and larval populations (Peter et al., 2005). Two groups of mosquito-borne diseases are particularly common: malaria and arboviral diseases; the second group of diseases includes yellow fever, dengue and various types of encephalitis (Stanier et al., 1983).

In many areas, particularly with less intensive transmission, environmental management can significantly reduce the spread of diseases. Water management in such locations should be based on a proper assessment and understanding of local vector ecology (Peter et al., 2005). The traditional common mosquito control technique has been chemical spraying, but other techniques have appeared during the last three decades, due to limitations of chemical spraying, such as the high cost of chemicals, mosquitoes' increased resistance to insecticides, the associated health risks to humans and domestic animals, and the disturbance of natural balances, such as the predator-prey relationship. Large, community-based mosquito control programmes are needed and should include educational and motivational activities for disrupting malaria transmission and preventing the destruction of the ecosystem (Yasuoka et al., 2006). A social-ecology approach is also needed in order to achieve permanent behavioural changes, such as the use of bed nets (Panter-Brick et al., 2006).

Schistosomiasis is another tropical disease, whose transmission cycle requires the contamination of surface waters by faeces and urine containing parasite eggs, the presence of snails and frequent contact with water. According to WHO, 200 million people are infected by this disease worldwide, leading to a loss of 1.58 million DALYs. Snail control with toxic molluscicides is expensive and logistically complex. It also affects other aquatic organisms, such as fish, and causes ecological and economic concerns. The snail population can be also reduced by canal lining, regular rapid draw-down of reservoirs and increased canal flow. Schistosomiasis could, in principle, be eliminated by behavioural changes, sanitation and safe water supply (Gryseels et al., 2006).

8.4 CHEMICAL RISKS AND THEIR CONTROL

Various types of chemicals occurring in urban waters may pose the risk of impairing human health. Two such groups, selected anthropogenic substances (primary pollution) and algal toxins (secondary pollution), are addressed in this section.

8.4.1 Anthropogenic compounds

The occurrence of various anthropogenic compounds in surface waters and groundwater has caused increasing concerns with respect to potentially adverse health and environmental effects. Municipal wastewater may contain harmful chemicals, including heavy metals, synthetic organic and inorganic chemicals, household chemicals, and pharmaceuticals and personal care products (PPCP). The grey water originating from kitchen sinks, bathrooms or laundries, can contain over 900 synthetic organic compounds, also referred to as xenobiotics (Erikson et al., 2002).

PPCPs have been studied widely during the last ten years and can be used as indicators or tracers of the human faecal pollution in the same way as the microbiological indictor organisms, such as *E. coli*. Almost 100 pharmaceuticals or their metabolites have been detected in sewage or the environment. Such compounds are carried from humans into the sewage treatment plant where they are partly biodegraded, partly adsorbed into sludge, and where some of the hydrophobic and non-biodegradable parts of these chemicals pass through the treatment processes and end up in receiving waters (Halling-Sørensen et al., 1998; Heberer, 2002).

A schematic presentation of the fate of PPCPs in surface water and groundwater is presented in Figure 8.1.

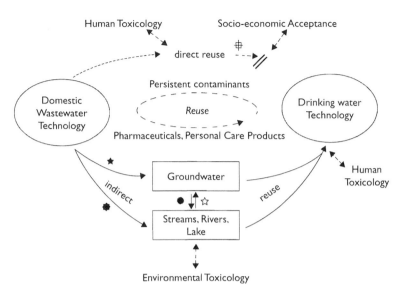

Figure 8.1 Fate of PPCPs in the urban hydrological cycle

Source: Poseidon 2006.

Even though some studies have found pharmaceuticals or their metabolites in drinking water, they are thought to be of a minor concern for human heath, because of their extremely low concentrations (low ng L^{-1}) compared to the therapeutic doses, which is in the range of milligrams (Lindqvist et al., 2005; Webb et al., 2003; Vieno et al., 2005).

Numerous researchers have investigated the occurrence and fate of pharmaceuticals in different environmental compartments trying to determine if sub- to low nanogram per litre concentrations are of concern with respect to aquatic organisms (Fent et al., 2006; Jones et al., 2004). Some impacts on aquatic organisms are reported, for example, for synthetic oestrogen, ethinyloestradiol, at concentrations as low as 0.1 ng/L (Purdom et al., 1994). In a comprehensive Swedish study, environmental hazard assessment and classification was done for twenty-seven representative pharmaceutical compounds (Carlsson et al., 2006a). The consumption data of selected pharmaceuticals were combined with local release scenarios in the Stockholm archipelago to produce predicted environmental concentrations (PEC) under very conservative conditions. Only the sex hormones oeastradiol and ethinyloestradiols can be considered to represent possible aquatic environmental risks in the concentrations that can occur in Swedish surface waters. More research is needed, however, to address such issues as the chronic ecotoxicological effects of PPCPs, their emissions and fate in treatment systems as well as in the environment (Carlson et al., 2006b).

The control measures for PPCPs include a substantial reduction in the use of pharmaceuticals to reduce the quantities entering the sewage, ecotoxicological labelling of personal care products containing perfumes and fragrances, and an effective treatment of municipal sewage (Poseidon, 2006).

The effective treatment of PPCPs in sewage systems follows the general principles of the removal of recalcitrant compounds, which includes the optimization of the removal of target compounds in the wastewater treatment plant, safe disposal of sludge and disposal of the treated effluents in water bodies with maximum dilution and attenuation. The low organic load, high sludge residence time and nitrification correlate positively with effective PCPP elimination and the removal of recalcitrant compounds in general (Carballa et al., 2004; Kreuzinger et al., 2004). Heavy rains and increased urban runoff into combined sewers reduce the wastewater temperature in cold climate, dilute wastewater and reduce the hydraulic residence time in wastewater treatment plants. This has resulted in a serious decrease of the PPCP removal efficiency (Vieno et al., 2007). Post-ozonation can be used to remove residual concentrations of PCPPs where receiving water bodies offer low dilution, or where the treated effluent is going to be injected into the soil (Ternes et al., 2003).

8.4.2 Algal toxins

Cyanobacteria (blue-green algae) have been reported in the freshwaters of over forty-five countries and in numerous brackish, coastal and marine environments. They grow intensively in eutrophic habitats, as a result of primary phosphorus and nitrogen loadings from point source pollution and runoff, and are thus considered to be a 'secondary pollution'. Toxicogenic Cyanobacteria are not listed among waterborne pathogens in the water industry, since they are unable to invade and grow in humans to cause disease. However, they are capable of causing adverse health effects by producing hepato- neuro- and cytotoxins, irritants and gastrointestinal toxins which can affect wild and domestic mammals, birds, amphibians, fish and human beings (Tarczynska et al., 2000; Mankiewicz et al., 2002). Human health effects are diverse: gastroenteritis, nausea, vomiting, fever, flu-like symptoms, sore throat, blistered mouth, ear and eye irritation, rashes, myalgia, abdominal pains, visual disturbances, kidney damage and liver damage. An increase in liver cancer incidence has been associated with exposure to cyanobacteria in raw water in China (Duy et al., 2000).

The exposure to algal toxins can take place via drinking water, but other exposure pathways also are possible. These include occupational or recreational direct skin contact to water, incidental or accidental ingestion of water, aspiration and inhalation of toxins. There is also some evidence of potential exposure via animal or vegetable foods. Some toxins can accumulate in fish and blue mussels. Some toxins and toxin-containing cells occur in spray irrigation water and in spray-irrigated salad lettuce (Codd et al., 1999). It is also possible that dairy cows, after oral exposure, secrete cyanobacterial toxins in their milk (Feiz et al., 2002).

The occurrence of toxic cyanobacteria in raw water reservoirs causes problems in the production of safe drinking water. The first task of risk management is the development of a guideline value (GV) for the concentration of toxins in drinking water. Guidance for establishing such GVs is available from WHO (2006).

In controlling algal toxins, prevention is better than cure. The long- and short-term catchment management, with respect to reducing external nutrients loading and eutrophication, is the most effective control option. Maintenance of landscape vegetation cover allows for the stabilization of water and matter flows and can diminish nutrient exports (see Chapters 4, 5 and 6). Other methods include the enhancement of self-purification and nutrient retention in river valleys or within the boundaries or aquatic habitats, based on a ecosensitive designs (see Chapter 4) and the rehabilitation of aquatic habitat structures (see Chapters 5 and 6). Ecosystem biotechnologies (Zalewski, 2002), ecohydrological measures (Wagner and Zalewski, 2000) and ecological engineering methods (Mitsch and Jorgensen, 2003), may additionally enhance efficiency of the reduction of the pollution export from the catchment.

When Cyanobacteria occur in raw water sources in the form of blooms of planktonic organisms, scum or mats, and biofilms of benthic and littoral species, there are still some options for risk control. By increasing public awareness of the dangers of and providing training to recognize cyanobacterial blooms, it is possible to prevent the use of the affected waters for recreational and fishing purposes.

Cyanobacterial growth may also be diminished by applying ecohydrological and engineering techniques and altering hydro-physiological conditions in water bodies, e.g., by shortening water retention time or increasing its flow or turbidity (Zalewski, 2002). Other short-term control options consist of the adjustment of water withdrawal depth, using bank filtration, or installing barriers to restrict the scum movement. The use of algaecides is also possible. The final stage of addressing the cyanobacteria problem is the removal of cell-bound or extracellular toxins from raw water by traditional or advanced water treatment processes. In water treatment, the target substances and organic and inorganic matrices are removed from the water phase either by separation or conversion. The most commonly studied algal toxins, such as microcystin and anatoxins, are water soluble. However, in many cases, toxins are found within the intact cells, and so their removal is based on the successful removal of algal matter without cell lysis by mechanical and chemical stress. Intact cells can be removed by microsieving, or by similar processes applied in conventional drinking water treatment trains with coagulation, clarification and sand filtration. However, such methods are ineffective for the removal of free extracellular toxins, which are best removed by adsorption onto granular or powdered activated carbon and oxidation. Ozonation has been found to be effective in completely destroying toxins (Hindberg et al., 1989; Keijola et al., 1988; Lepistö, 1994; Jurczak et al., 2005).

8.4.3 Sewage sludge use as an option for the control of chemical risks

In sewage treatment, suspended, colloidal and dissolved organic and inorganic matter is transferred into sludge, which can be separated from the aqueous phase, the treated effluent. Sludge is rich with organic humic-like matter and nutrients. Typical stabilized wastewater sludge contains 5% nitrogen, 2.3% phosphorus and 0.3% potassium. In principle, sewage sludge, also called biosolids, is a valuable resource that should be recycled back into agricultural soils as a source of nutrients and organic matter. The major problem with using sewage sludge as a fertilizer is the possible presence of chemical and microbiological pollutants and nitrogen. Municipal wastewater contains PPCPs, heavy metals and hydrocarbons from runoff and industrial sources. During the primary and secondary treatment (pre-clarification and biological treatment) processes, hydrophobic organic impurities and heavy metals become adsorbed onto sludge. The problematic heavy metals are cadmium, chrome, copper, mercury, nickel, lead and zinc. Furthermore, Harrison et al. (2006) found 516 different organic compounds in sewage sludge, and the commonly found compounds are also reported by the EC (2001). Pathogens of concern include *E. coli*, Salmonella, Enteroviruses, Giardia and some helminths.

The discussion of using sewage sludge in EU countries as a fertilizer has been going on since the beginning of the 1990s. This process is regulated by the European Union Council Directive (EEC 86/278), with a new version under preparation. Various stakeholders use different approaches, and sludge disposal practices greatly vary. For example, Sweden has banned the use of sewage sludge in agriculture. Sewage sludge may be also used as fertilizer or in short crop forestry (e.g., energetic willow plantations) for the production of biomass. Biomass can be used as bioenergy, contributing to the economic equation of wastewater treatment plants, as well as decreasing CO_2 emissions (Wagner-Lotkowska et al., 2004; Zalewski and Wagner, 2006). Such an example is presented in more detail in Chapter 9, presenting a system approach to restoring cycles of matter and pollutants in the City of Lodz, Poland. Sewage sludge can be also used in forestry or for fertilizing and improving the structure of soils in urban green areas, thus improving vegetation growth and the area's aesthetic appeal. The current practices of sludge disposal in developed countries are strongly opposed to the closing-the-loop approach, which is emphasized strongly in ecological sanitation.

8.5 DROWNING AND OTHER WATER-RELATED RISKS OF AQUATIC HABITATS

About 80% of all natural disasters between 1996 and 2005 were of meteorological or hydrological origin. Extreme hydrological events interact with other water-related hazards, such as chemical spills, aquifer depletion, land subsidence, salination of arable land and marine intrusion (World Water Assessment Programme, 2006). There have been a number of fluvial floods resulting in massive losses of human life: the Yellow River in China (nearly 1 million people) and Bangladesh (139,000 people) (WHO, 2001).

The mortality caused by drowning is equivalent to cases of poisonings, falls and fires, in terms of numbers. To minimize risk, the World Heath Organization (WHO) has launched guidelines for safe recreational water environment, which also includes the prevention of drowning (WHO, 2004a). Most drown victims are children under the age of fifteen, and 97 percent of the cases occur in low- or middle-income countries (Peden and McGee, 2003). Drowning is the third leading cause of deaths in children aged 1 to 5, and it is the leading cause of mortality due to injury. This risk is not related only to recreational water use in freshwater or tidal lagoons. In less developed countries, loss of human life due to natural disasters and the occupational drowning of sailors and fishermen is common.

It has been estimated that 80% of these cases could be prevented by such measures as public education regarding hazards and safe behaviour, the continual adult supervision of children, restriction of alcohol use, provision of rescue services, development of rescue and resuscitation skills among general public, and the compulsory use of lifejackets when boating (WHO, 2003).

8.6 INTEGRATED WATER RISK MANAGEMENT

8.6.1 General

Preventing problems is always better than remediating them once they have occured. Success in the reduction or elimination of health risks depends on the means applied by public health agencies to control, contain and mitigate health hazards causing risks. A successful risk reduction strategy should adopt an integrated approach combining such disciplines as ecology, engineering, management, public health, chemistry and agriculture. Health risk reduction is not possible without risk identification, assessment and control, which involves regulating, monitoring and enforcement measures.

The health risk depends on the type of the agent of concern. Microbial pathogens can cause acute impacts on human health. On the other hand, potentially toxic chemicals occurring at low levels exert effects only after long-term exposure.

Drinking water safety issues involve various stakeholders, such as consumers and experts (researchers, regulators and suppliers) from such disciplines as engineering, chemistry, public health, economics and law. Risk perceptions vary significantly among stakeholders and are not based solely on quantifiable scientific data. Quite often, unfamiliar, new and coerced risks are considered more significant than the old and familiar risks under control. In general, openness in communication and transparency in the decision-making process, as well as the possibility to participate in problem resolution, increase the satisfaction of water consumers. Among stakeholders, scientists are trusted more than water suppliers (Baggett et al., 2006).

To avoid human health problems caused by wastewater, the 'multi-barrier approach' is applied by introducing barriers between the sewage and the population (Toze, 2006). Controls should be applied as close to the source of the risk as possible. In practice, this approach implements the following basic idea: if a single natural or technical process controlling harmful agents becomes ineffective, the contamination of drinking water by such agents will be avoided by other measures following afterwards. Furthermore, the specific removal potentials of specific steps in the multi-barrier system support the efficiency and stability of the whole system by achieving efficient and safe water supply; minimizing environmental impacts of wastewater; and avoiding potential human health hazards. With respect to microbiological risks, even short barrier failures can lead to an unacceptable outcome – an outbreak of a waterborne epidemic disease (Hamilton et al., 2006).

Systematic screening of the source and composition of wastewater and possible drinking water treatment prior to use can provide an estimate of the quality and magnitude of possible risks. New tools have been introduced, particularly by WHO, to assess and control the risks related to drinking water. Water Safety Plans (WSP) and the Hazard Assessment and Critical Control Point approach are powerful tools for risk reduction. When the most important hazards are known, the resources available for risk reduction can be allocated in the most cost-efficient way.

8.6.2 Water Safety Plans (WSP)

Water Safety Plans serve as comprehensive risk assessment and risk management tools for maintaining safe and secure water supply, from the catchment sources to the consumer. The value of WSPs is recognized in a recent version of the WHO drinking water guidelines (WHO, 2004b). Water safety plans provide a framework, or a road map, for establishing preventive, step-by-step processes for water risk management. The key idea is to create an integrated approach to water safety across the four stages of water supply: catchment, treatment, distribution and the customer's plumbing system. The risks for drinking water contamination can arise at many points from the catchment all the way to the tap. For example, a catchment with urban areas and urban aquatic habitats (streams, rivers, lakes, and reservoirs) may contain many sources of chemical and microbiological contamination from urban runoff, septic tanks, sewage treatment plants, pets, wildlife and livestock. The raw water source itself can be affected by algal blooms or release of toxic compounds from sediments (e.g., due to anaerobic conditions or mechanical mixing). The quality and quantity of raw water can change quickly, which puts pressure on drinking water treatment plants. Drinking water treatment has to produce safe water in all conditions, not just most of the time. The protection of water quality in the distribution system is the final stage of the multi-barrier approach to safe drinking water supply. Several pathways exist for contaminants to enter the distributing system: through storage facilities, water towers, cross-connections, water main repairs and intrusion through points of leakage due to the pressure fluctuation. Drinking water should be ideally non-corrosive, biochemically stabile and non-scaling; in reality, this ideal is hardly ever reached. It is particularly difficult to keep the distribution system clean as infrastructure ages. Water utilities need to continually evaluate the options of rehabilitation versus replacement, for example based on pipes' structural condition, leakage volume and the disruption of services to customers.

Water safety plans comprise at least three essential actions:

- System assessment to determine whether the drinking-water supply chain can deliver water that meets health-based targets
- Identifying control measures that will control identified risks and ensure that the health-based targets are met
- Management plans describing actions to be taken during normal or incident conditions and documentation of the system assessment, monitoring, communication plan and supporting programme.

8.6.3 Hazard Analysis and Critical Control Point (HACCP)

Most water safety plans are based on the adaptation of the HACCP procedure, which was developed by the US Space Agency in the 1960s. It has been used to monitor the hazards for protection of water supplies in recent times, particularly in Australia, Austria and Iceland. It shifts emphasis from resource-intensive end-product inspections and testing to preventive controls of hazards at all stages of water supply system. This involves systematic identification, evaluation and continuous record keeping and control. Necessary preventive methods are defined with respect to preventing, eliminating or at least reducing hazards to an acceptable level. For protection against pathogens, numerical limits are not useful since the sampling and analysis process is too slow. For pathogens, the barriers include protection measures in the catchment, storage ahead of treatment and the treatment stages.

It is of interest to examine the recently released FAO guideline for HACCP application in the food industry (CAC, 1997), which is also relevant to the water industry. It comprises seven key steps:

1. perform a hazard analysis
2. determine the critical points (CCP)
3. establish one or several critical limits

4. establish a CCP monitoring system
5. establish corrective action to be taken if monitoring indicates that a specific CCP is no longer under control
6. establish a procedure for verification confirming successful operation of the HACCP system
7. introduce a documentation system taking into account all processes and records, according to their principles and their applications.

Some essential terms used in this approach are defined below.

Hazard: all biological, chemical or physical characteristics that could cause an unacceptable health risk for the consumer.
Risk: the assessment of the possible occurrence of danger or hazard.
HACCP System: recognized international standard for comprehensive type of quality control. The basis for this control system is a risk analysis of all production and distribution installations.
CCP: Critical Control Point, which is defined as every point or process in the entire water supply system, where the loss of control would cause an unacceptable health risk.
Hazard analysis: comprises analysis of hazards and identification of the associated control points. Hazards can be microbiological, chemical, physical or radioactive parameters whose presence impacts on the consumer's health. The causes of introduction of hazards can be accidental, uncontrolled events, vandalism, sabotage (terrorism), natural disasters or operational parameters.

The HACCP procedure is conducted by a team of specialists from all the fields involved and all operational levels of water management. HACCP is based on answering three questions: What is the hazard? How is the hazard fixed? and How do you know that the hazard has been fixed?

8.7 CONCLUSIONS

Risks threatening the balance of ecosystems also affect human well-being by endangering the chemical and microbiological quality of water resources. The uncontrolled growth of residential, industrial and agricultural activities has a substantial impact on the flux and quality of water that recharges aquifers and discharges into streams, lakes and wetlands. Contamination sources may vary in scope and composition, but they affect the urban water sustainability and public health regardless of the level of development. Perhaps the most important finding of the discussion in this chapter is the fact that while people derive tremendous benefits from good quality aquatic habitats, and for this very reason many human settlements are located on the shores of such water bodies, degraded habitats increase hazards and risks connected with the uses of such habitats for water supply, drainage and flood protection, conveyance/disposal of treated urban wastewaters, and ecological functions. Thus, a consistent application of ecosensitive approaches to urban aquatic habitats is beneficial not only to the habitats themselves and the fauna and flora they support, but also for urban populations benefiting from such water bodies in a number of ways, including ecosystem services and better health.

REFERENCES

Baggett, S., Jeffrey, P. and Jefferson, B. 2006. Risk perception in participatory planning for water reuse, *Desalination*, Vol. 187, pp.149–58.

Betancourt, W.Q. and Rose, J.B. 2004. Drinking water treatment process for removal of Cryptosporidium and Giardia, *Veterinary Parasitology*, Vol. 126, pp. 219–34.

Bixio, D., Thoeye, C., De Konig, J., Joksimovic, Savic, D., Wintgens, T. and Melin, T. 2006. Wastewater reuse in Europe, *Desalination*, Vol. 187, pp. 89–101.

CAC (Codex Alimentarius Commission) Hazard analysis and criteria control point (HACCP) system and guideline for its application, Annex to CAC/RCP 1-1969, Rev 3; 1997, Available: http://www.fao.org/documents (last accessed January 2007).

Cairns, J. 1999. Exemptionalism vs environmentalism: the crucial debate on the value of ecosystem heath. *Aquatic ecosystem Heath & Management*, Vol. 2. pp. 331–38.

Carballa, M., Omil, F., Lema, J.M., Llombard, M., Garsia-Jares, C., Rodrigues,I., Gomez, M. and Ternes, T. 2004. Behaviour of pharmaceuticals, cosmetics and hormones in a sewage treatment plant, *Water Research*, Vol. 38, pp. 2918–26.

Carducci, A., Tozzi, E., Rubelotta, E., Casini, B., Cautini, L. Rovini, E., Muscillo, M., and Pacini, R. 2000. Assessing airborne bacterial hazards from urban wastewater treatment, *Water Research*, Vol. 34. pp. 1173–78.

Carlsson, C., Johansson, A-K., Alvan, G., Bergman, K. and Kűhler, T. 2006a. Are pharmaceuticals potent environmental pollutants? Part I: Environmental risk assessment of selected active pharmaceutical ingredients. *Sci Total Environment*, Vol. pp. 364, 367–87.

Carlsson, C., Johansson, A-K., Alvan, G., Bergman, K. and Kűhler, T. 2006b. Are pharmaceuticals potent environmental pollutants? Part II: Environmental risk assessment of selected pharmaceutical excipients. *Sci Total Environment*, Vol., pp. 364, 388–95.

Codd, G.A., Medcalf, J.S. and Beattie, K.A., 1999, Retention of *Microcystis aeruginosa* and microcystin in salad lettuce after spray irrigation with water containing cynaobacteria. *Toxicon*, Vol. 37, pp. 1181–86

Duy, T.N., Lam, P.K.S., Shaw, G.R. and Connell, D.W. 2000. Toxicology and risk assessment of freshwater cyanobacterial (blue-green algal) toxins in water. *Rev. Environ. Contam. Toxicol.* Vol. 163, pp. 113–86.

EEC 86/278. Council Directive on the protection of the environment, and in particularly the soil, when sewage sludge is used in agriculture. Available: http://eur-lex.europa/LexUriServ.do?uri=CELEX:31986L0278:EN:HTML (last accessed January 2007)

Erikson, E., Auffarth, K., Henze, M. and Ledin, A. 2002 Characteristics of grey wastewater, *Urban Water*, Vol. 4, pp. 85–1004.

European Commission, 2001. *Disposal and recycling routes for sewage sludge. Part 3 – Scientific and technical report* http://ec.europa.au/environment/waste/sludge/pdf/sludge_disposal3.pdf

EC 2006/7. Directive of the European Parliament and of the Council of 15 February 2006 concerning the management of bathing water quality and repealing Directive 76/160/EEC Available: http://europa.eu/scadplus/leg/en/lvb/l28007.htm (assessed January 2007).

Feiz, A.J., Lukondeh, T., Moffitt, M., Burns, B.P., Naidoo, D., Vedova, J.D., Gooden, J.M. and Neilan, B.A., 2002. Absence of detectable levels of cyanobactrial toxins (microcystin-LR) carry-over into milk. *Toxicon*, Vol. 40, pp. 1173–80.

Fent, K., Weston, A.A. and Caminada, D. 2006. Ecotoxicology of human pharmaceuticals. *Aquatic Toxicology*, 76, pp. 122–59.

Fracchia, L., Pietronave, S., Rinaldi, M. and Martinotti, M.G. 2006. Site-related airborne biological hazard and seasonal variations in two wastewater treatment plants, *Water Research*, Vol. 40, pp. 1985–94.

Georgiou, S. and Batman, I.J. 2005. Revision of EU Bathing Water Directive: economic costs and benefits. *Marine Pollution Bulletin*, Vol. 50. pp. 430–38.

Gryseels, B., Polman, K., Clerinx, J. and Kestens, L., 2006. Human schistosomiasis, *The Lancet*, Vol. 368, Sept. 23, pp. 1106–18.

Hallig-Sørensen, B. Nors Nielsen, S., Lanzky, P.E., Ingerslev, F., Holten Litzhof, H.C and Jorgensen, S.E., 1998. Occurrence, fate and effects of pharmaceutical substances in the environment – a review, *Chemosphere*, 36 (2), pp. 357–93.

Hamilton, P.D., Gale, P. and Pollard, S.J.T., 2006. A commentary on recent water safety initiatives in the context of water utility risk management, *Environmental International*, Vol. 32, pp. 958–66.

Harrison, E.C., Oakes, S.R., Hysell, M and Hay, A. 2006. Organic chemicals in sewage sludge – Review, *Science of Total Environment*, Vol. 367, pp. 481–97.

Heberer, T. 2002. Occurrence, fate and removal of pharmaceutical residues in the aquatic environment: a review of recent research data. *Toxicology Letters*, 131, pp. 5–17.

Hindberg, K., Keijola, A., Hiisivirta, L., Pyysalo, H. and Sivonen, K. 1989. The effect of water treatment processes on the removal of *Microcystis* and *Oscillatoria* cyanobacteria: a laboratory study. *Water Research*, Vol. 23. pp. 979–84.

Hunter, P.R. and Thompson, R.C.A., 2005. The zoonotic transmission of Giardia and Cryptosporidium, *International Journal of Parasitology*, Vol. 35, pp. 1181–90.

Jacks, G., Forsberg, J., Magoub, F. and Palmqvist K. 2000, Sustainability of local water supply and sewage system – a case study in a vulnerable environment. *Ecological Engineering*, Vol. 15, pp.147–53.

Jones, O.A.H., Voulvoulis, N. and Lester, J.N. 2004. Potential ecological and human health risks associated with the presence of pharmaceutically active compounds in the aquatic environment. *Critical Reviews in Toxicology*, 34(3), pp.335–50.

Jurczak, T., Tarczynska, M., Izydorczyk, K., Mankiewicz, J., Zalewski, M. and Meriluoto, J. 2005. Elimination of microcystins by water treatment processes – example form Sulejow Reservoir, Poland, *Water Research*. Vol. 39, pp. 2394–406.

Mankiewicz, J., Walter, Z., Tarczyńska, M., Palyvoda, O., Wojtysiak-Staniaszczyk, M., Zalewski, M. 2002. Genotoxicity of cyanobacterial extracts with microcystins from Polish Water Reservoirs as determined by the SOS Chromotest and comet assay. *Environmental Toxicology* 17(4): pp. 341–50.

Karim, M.R., Manshadi, F.D., Karpiscak, M.M. and Gerba, C.P. 2004. The persistence and removal of enteric pathogens in constructed wetlands. *Water Research Volume*: 38, Issue: 7, pp. 1831–37.

Keijola, A.M., Hindberg, K., Esala, A., Sivonen, K. and Hiisivirta, L. 1988. Removal of cyanobacterial toxins in water treatment processes: a laboratory and pilot-scale experiment. *Tox. Assess*. Vol 3. pp. 643–56.

Kreuzinger, N., Clara, M., Strenn, B. and Kroiss, H. 2004. Relevance of the sludge retention time (SRT) as design criteria for wastewater treatment plant for the removal of endocrine disrupters and pharmaceuticals from wastewater, *Wat. Sci. Technol*. Vol. 50, pp.149–56.

Lepistö, T. Lahti, K., Niemi, J. and Fardig, M. 1994. Removal of cyanobacterial and phytoplankton in four Finnish waterworks. *Arch. Hydrobiol. Algological Studies*, Vol. 3, pp.167–81.

Lindqvist, N., Tuhkanen, T. and Kronberg, L. 2005. Occurrence of acidic pharmaceuticals in raw and treated sewages and in receiving waters. *Water Research*. Vol. 39, 2219–28.

Panter-Brick, C., Clarke, S.E., Lomas, H., Pinder, M. and Lindsay, S.W. 2006. Culturally compelling strategies for behaviour change: A social ecology model and case study in malaria prevention, *Social Science & Medicine*, Vol. 62, pp. 2810–25.

Peden, M.M. and McGee, K. 2003. The epidemiology of drowning worldwide, *Injury Control and Safety Promotion*, Vol. 10, pp.195–99.

Peter, R.J., van den Bossche, P., Penzhorn, B.L. and Sharp, B. 2005. Tick, fly and mosquito control – Lesson from the past, solutions for the future, *Veterinary Parasitology*, vol. 132, pp. 205–15.

Poseidon 2006. Assessment of technologies for the removal of pharmaceuticals and personal care products in sewage and drinking water facilities to improve the indirect and direct potable water reuse. (http://ec.europa.eu/research/endocrine/pdf/poseidon_final_report_ summary_en.pd. (Accessed December.2006).

Purdom, C.E., Hardiman, P.A., Bye, V.J., Eno, N.C., Tyler, C. and Sumpter, J. 1994. Estrogenic effects of effluents from sewage treatment works, *Chem Ecol* Vol 8, pp.275–85.

Sahu, A., Grimberg, S.J and Holsen, T.M. 2005. A static water surface sampler to measure bioaerosol deposition and characterize microbial community diversity, *Journal of Aerosol Science*, Vol. 36, pp. 639–50.

Scandura, J.E. and Sobsey, M.D. 1997. Viral and bacterial contamination of groundwater from on-cite sewage treatment system, *Water Science and Technology*, Vol. 35, pp.141–46.

Schönning, C. and Stentröm, T.A. 2004. Guidelines on the Safe Use of Urine and Faeces in Ecological Sanitation System, *Stockholm Environment Institute*, Report 2004-1, http//www. ecosanres.org (last accessed August 2006).

Stanier, R.Y., Adelberg, E.A. and Ingrham, J.L. 1983. *General Microbiology* 4th edn. Macmillan.

Szabo, H.M. 2004. The effect of small scale wastewater disposal on the quality of well water, *M.Sc thesis*, Tampere University of Technology.

Tambo, N. 2002, A new water metabolic system, Water 21, December, pp. 67–68.

Tarczyńska, M., Nałęcz-Jawecki, G., Brzychcy, M., Zalewski, M. and Sawicki, J. 2000. The toxicity of cyanobacterial blooms as determined by microbiotests and mouse assay. G. Personne et al. (eds) *New Microbiotest for Routine Toxicity Screening and Biomonitoring*. Kluwer Academic/Plenum Publishers, New York, pp. 527–32.

Ternes, T., Stuber, J., Hermann, N., McDowell, D., Ried, A., Kampmann, M. and Teiser, B. 2003. Ozonation: A tool for removal pf pharmaceuticals, contrast media and musk fragrances from Wastewater? *Water Research*, Vol. 37, pp. 1976–82.

Toze, S. 2006. Water reuse and heath risks – real vs. perceived, *Desalination*, Vol. 187, pp. 41–51.

van Cuyk, S, Siegrist, R., Logan, A. Masson, S., Fisher, E. and Figuroa, L. 2001. Hydraulic and purification behaviours and their interactions during wastewater treatment in soil infiltration systems, *Water Research*, Vol. 35, pp. 953–64.

Vieno, N., Tuhkanen, T. and Kronberg, L., 2005. Seasonal variation in the occurrence of pharmaceuticals in effluents from a sewage treatment plant and in the recipient water, *Environmental Science and Technology*, 39(21) 8220–26.

Vieno, N., Tuhkanen, T. and Kronberg, L. 2007. Elimination of Pharmaceutical in sewage treatment plants in Finland, *Water Research*, Vol. 41, pp. 1001–12

Wagner, I. and Zalewski, M. 2000. Effect of hydrological patterns of tributaries on biotic processes in a lowland reservoir – consequences for restoration, Ecological Engineering, Vol. 16, pp. 79–90.

Wagner-Lotkowska, I., Bocian, J., Pypaert, P., Santiago-Fandino, V. and Zalewski, M. 2004. Environment and economy – dual benefit of ecohydrology and phytotechnology in water resources management: Pilica River Demonstration Project under the auspices of UNESCO and UNEP. *Ecology and Hydrobiology* Vol. 4, No. 3, pp. 345–52.

Williams, J., Bahgat, M., May, E., Ford, M. and Butler, J. 1995. Mineralisation and pathogen removal in gravel bed hydroponic constructed wetlands for wastewater treatment. *Water Science and Technology*, Vol. 32, Issue: 3, pp. 49–58

Webb, S., Ternes, T. Gilbert, M. and Olejniczak, K. 2003, Indirect human exposure to pharmaceuticals via drinking water, *Toxicology Letters*, Vol. 142, pp.157–67.

Taking care, Geneva. A available online: http://www.who.int/water_sanitation_health/ takingcharge/en/ (last accessed 21.1.2007).

UN. 2005. Millennium Ecosystem Assessment

WHO. 2001. Water for heath – WHO statistics annex: Deaths by causes, sex, mortality stratum in WHO regions estimated 2002, in http://www.who.int/whr/2004/annex/topic/en/annex_2_en.pdf (last accessed December 2006).

WHO. 2003. Guidelines for safe recreational water environments. Volume 1: Coastal and fresh waters: available online: (last accessed 22.2.2007) http://www.who.int/water_sanitation_health/ bathing/srwe1execsum/en/index.html.

WHO. 2004a. Guideline for safe recreational water environments, Volume 1. Coastal and fresh waters. Available online: http://www.who.int/water_sanitation_health/bathing/srwe1execsum/en/ (last accessed December 2006).

WHO. 2004b. Guidelines for drinking water quality. 3rd. edition, Geneva. Available at http://www.who.int/water_sanitation_health/dwq/en (last accessed January 2007).

WHO. 2006. Guidelines for drinking-water quality, third edition, incorporating first addendum: Chapter 7: Microbial aspects http://www.who.int/water_sanitation_health/dwq/gdwq0506_7.pdf (last accessed August.2006).

World Health Report. 2004. http://www.who.int/whr/2004 (last accessed December 2006).

World Water Assessment Programme. 2006. *Water a shared responsibility, The United Nations World Water Development Report 2*, UNESCO, Paris.

Zalewski, M. (ed.) 2002. Guidelines for the Integrated Management of the Watershed. Phytotechnology and Ecohydrology. UNEP, Division of Technology, Industry and Economics. Freshwater Management Series No.5.

Zalewski, M. and Wagner, I. 2006. Ecohydrology – the use of water and ecosystem processes for healthy urban environments. Aquatic Habitats in Integrated Urban Water Management. *Ecohydrology & Hydrobiology*. Vol. 5. No 4, pp. 263–68

Yasuoka, J., Levins, R., Mangione, T.W. and Spielman, A. 2006. Community-based rice ecosystem management for suppressing vector anophelines in Sri Lanka. *Transitions of the Royal Society of tropical medicine and Hygiene*, Vol. 100, pp. 995–1006.

Chapter 9

Integrated management of urban aquatic habitats to enhance quality of life and environment in cities: Selected case studies

Azime TEZER

Urban and Regional Planning Department, Istanbul Technical University (ITU), Taskisla 34437, Taksim, Istanbul, TURKEY

9.1 INTRODUCTION

Aquatic habitats, and water resources in general, have attracted human settlements since the earliest times. Over centuries and millennia, many such settlements evolved into towns, cities and urban agglomerations that impose severe threats to water resources and their sustainability. A purely technical approach to water resources management, which has predominated in many cities for a long time, supported smaller populations and prevented visible environmental changes. However, considering the enormous acceleration of rural population migration to cities during the recent decades (UN, 2004), population growth and high rates of urbanization, the conventional technical approach can no longer cope with problems of urban waters and is not affordable in most, if not all, developing countries. Further technological manipulations of water systems, without consideration of aquatic habitats' features and needs, may not only affect the structure and functioning of ecosystems, but might also further increase their vulnerability. An impaired ability to provide ecosystem services also impacts economic, social and cultural aspects of life in cities.

A growing awareness of this situation has initiated a promising multidisciplinary and multi-sectoral discussion of the sustainability of cities and their water resources. It urges planners, engineers, scientists, economists, social scientists, political decision-makers and other stakeholders to work together for better practices that would reconcile and mitigate urban pressures on aquatic habitats and maximise their ability to better serve the needs of urban populations. Urban water management plans often and increasingly integrate urban development and aquatic habitats, no longer considering habitats as passive targets for restoration or protection. Instead, aquatic habitats are now becoming an active element of efficient management – playing a role in water quality improvement, stabilizing the catchment hydrological cycle, mitigating floods, improving microclimate and human health, and bringing other benefits. After all, 'the battle for sustainable development will be won or lost in cities' (UNCHS, 1996).

The selected case studies (CS) presented in this chapter give examples of such modern planning experiences in eight cities of the world where the process of innovative planning has already started and is currently in various stages of advancement – starting from an early conceptual planning stage (e.g., the cities of Phoenix, US and Istanbul, Turkey), through the first phases of implementation. They range from an environmental assessment and demand-led research (Vienna, Austria, Lyon, France and the Adige region in Italy), to already ongoing implementation giving some early positive results (Cape Town, South Africa; Sharjah, United Arab Emirates and Lodz, Poland). Regardless of the stage of advancement, all case studies break the continuous and alarming trend of unsustainable urbanization and show the importance and necessity of exploring various ways of more consciously considering aquatic habitats and harmonizing technological measures with ecosensitive approaches in integrated urban water management (IUWM).

The objective of this chapter is to give examples and learn from different CSs how aquatic habitats can be addressed in urban water management plans. As water-related problems are believed to increase exponentially, similar to population growth, these case studies address various urban development dynamics and populations (examining developments in cities with populations between 700,00 and 10 million inhabitants), problems and priorities. The scale and character of the addressed issues determine the spatial scale of implementation (regional, urban, catchment scale and local), showing the potential for different approaches addressing well defined priorities. For example, the case study of the Adige River in Northern Italy is an example of a regional-scale project addressing several cities along the valley of the major watercourse in the catchment of a tourist region of Italy. Some other CSs address aquatic habitats of high importance for meeting water-related services in metropolitan areas, such as demands for surface water in a rapidly growing city in a hot, dry desert area (Phoenix, US), flood protection of a historical city (the Danube in Vienna) or stormwater and combined sewer overflow management (the Yzeron River in Lyon and the Sokolowka and Ner rivers in Lodz).

These case studies address a wide range of climatic conditions, aquatic habitat types (e.g., large and small rivers, reservoirs and ponds, associated terrestrial ecosystems) and their roles for urban residents. There are examples of transitional land and water urban aquatic habitats, such as floodplains and wetlands (e.g., the Lobau Biosphere Reserve in Vienna, the Moddergat River in Cape Town and the Wasit Nature Reserve in the United Arab Emirates), which are highly important for the improvement of water quality, the mitigation of the catchment impacts including hydrological extreme events, and the stabilization of biodiversity and landscape resilience. Some of the other featured CSs strengthen the idea of better land-use control and urban development planning for achieving sustainable aquatic habitats (e.g., the Adige River and the Omerli Watershed in Istanbul). They show that the inclusion of ecological units into the urban planning process and the coordination of actions of the related stakeholders and communities with activities of a multidisciplinary group of professionals is a necessity.

As mentioned in Chapter 2, UAH management requires various approaches, ranging from preservation to restoration and from rehabilitation to remediation. Their selection depends on the degree of habitat modification, and a range of such habitats is also covered by the presented CSs. The Lobau floodplain (UNESCO Biosphere Reserve), the Adige River (including the Natura 2000 sites) and the Wasit Nature Reserve give examples of land-use control measures applied towards the preservation of some highly valuable areas protected by national legislation and international conventions or declarations. Another example is the Omerli Watershed, which represents an exceptionally high valued area whose protection is suggested here with the belief that this watershed could be included among world environmental heritage sites.

Some highly degraded areas and aquatic habitats are represented by the case studies of the Moddergat, Yzeron, and the Sokolowka and Ner rivers. They are examples of habitats that should undergo rehabilitative/restorative measures. However, considering the magnitude of impacts on these streams and their locations within urban areas, they require innovative methods based on remediation (e.g., phytotechnology) and the enhancement of bio-assimilative and absorbing capacities of ecosystems (e.g., ecohydrology). The variety of urban impacts superimposed on specific aquatic habitats usually requires a combination of all the above approaches, in addition to their careful integration into the urban water management, based on an in-depth analysis of local constrains, opportunities, needs and priorities. Such an attitude, together with the strengthening of cooperation among decision-makers, researches and the public, will certainly make for healthier ecosystems, better quality of life in urban settlements, lower risks of water-related hazards, and better interactions between nature and society in general.

9.2 MODDERGAT RIVER REHABILITATION AND FLOOD MANAGEMENT PROJECT CAPE TOWN, SOUTH AFRICA

Elizabeth DAY

University of Cape Town, Freshwater Research Unit, Rondebosch, 7700, South Africa

The Moddergat River is located on the urban edge of Cape Town, South Africa, in the Cape Floral Kingdom. The city has a population of around 3.5 million and a Mediterranean climate. Under natural conditions, the river would have been a diffuse, seasonal wetland. Major impacts on the system include channelization, urban and agricultural runoff, invasive terrestrial and wetland species, changes in natural hydrological conditions and a poor water quality. Improvements and rehabilitation measures addressed flood control, public safety and security, amenity and ecological functioning. Major actions implemented were soft engineering measures in the riverine area (gabion weirs, landscaping and replanting) and the widening of the floodplain (see Figure 9.1).

9.2.1 Background

The residential areas of Firgrove and Macassar on the Cape Flats, Cape Town, are prone to flooding by the Moddergat River. This is largely the result of alterations in

Figure 9.1 The Moddergat River (See also colour plate 7)
Photo credit: E. Day.

natural channel morphology, encroachment by buildings and other structures into the floodplain, and constriction of the channel by culverts. An initial hydrological study recommended hard canalization as the most suitable engineering option to protect affected areas from flooding. However, as well as being associated with prohibitive costs, canalization is also less popular because of its negative ecological and aesthetic characteristics. A revised, softer treatment of the stormwater problem was devised by a multi-disciplinary team.

9.2.2 Key aquatic habitat issues in urban water management

Prior to the implementation of the present project, the Moddergat River was highly modified from its natural condition, which is thought to have been a wide, braided wetland-associated system, meandering through soft substrata. At the beginning of the twentieth century, the river and wetland swathe was diverted some distance to the west of its natural course and confined to a single channel.

At the start of the project, the river in the study area (see Figure 9.2) was impacted by the following factors:

- encroachment of buildings and infilling into the floodplain
- extensive erosion,
- channelization, including straightening of the channel and steepening of banks

Figure 9.2 Moddergat River before intervention, showing deep, steep-sided channel in section recently dredged of reeds (See also colour plate 8)

Note: Litter on channel edges, lack of connectivity between aquatic and terrestrial habitat; prevalence of invasive kikuyu grass (*Pennisetum clandestinum*) along the channel margins.
Photo credit: CCA Environmental.

- loss of virtually all indigenous riparian vegetation
- encroachment of alien vegetation into the river margins
- nutrient-enrichment, and elevated concentrations of faecal coliform bacteria (CCA Environmental, 1999a; Day and Ractliffe, 2002; Day et al., 2005).

Moreover, dense growth of the invasive bulrush *Typha capensis* in places along the channel constituted both a safety hazard (the channel was hard to climb out of) and a security threat (reedbeds in this area often harbour criminal elements and facilitate attacks on passers-by).

9.2.3 Objectives of the case study and method of implementation

The primary objective of the project was to address flood control, by accommodating the 1:50 year flood; with secondary objectives including improving the amenity value along the river, preventing erosion of the riverbed and banks and, where possible, improving ecological function.

The project design involved the use of 'soft' engineering methods and included the following aspects:

- A macro-channel with a capacity designed to accommodate the estimated 1:50 year flood volume, with allowance for colonization of the low-flow channel section by dense reed beds
- A low-gradient trapezoidal channel, in which the low-flow channel was lined with loose river boulders
- Installation of gabion weir energy dissipaters at intervals along the channel
- Landscaping and planting (mainly grassing) of the flood channel, so as to create an attractive and safe amenity, useful for playing and walking
- Planting the low-flow channel edges with hardy, indigenous plant species
- Safeguarding against erosion, particularly in the early stages of the project before the establishment of vegetation, by laying a high density polyethylene mesh along the inner edge of the high-flow channel, effectively protecting the intersection between the low- and high-flow channels (CCA Environmental, 1999b).

The low-flow channel was constructed as a flat-bottomed base, lined and edged with loose river cobbles (see Figure 9.3). The design does not allow for maintenance activities like dredging sediment, and it was intended for sediment to settle on the river bed. Sedimentation occurred rapidly during the first year, partially covering sections of the boulders. Limited planting was carried out in sedimented areas.

Five gabion weirs, with scour-protective Reno mattress were constructed in different sections of the river, where they acted as energy dissipaters (see Figure 9.1). Gabions were also installed in steep sections of the high-flow channel (especially along bends in the river course), where spatial constraints prevented adequate widening and sloping of upper banks.

The high-flow channel (see Figure 9.4) had a flattened base, elevated some 0.5 m above the low-flow channel and edged, on the water side, by piled river cobbles, inclining up into steeply sloped banks (ca 1:3 to 1:4). In areas where open space was

Figure 9.3 Cobble-lined low-flow channel, with vegetation established along the edge soon after completion of channel. (Rapid invasion of the channel by reeds results in the long-term creation of a dense, reeded low-flow channel – this could result in improved water quality in the main channel; Photo credit: E. Day) (See also colour plate 9)

Figure 9.4 View of grassed floodplain and densely reeded low-flow channel, four years after completion. (Dense reedbeds reduce in-stream habitat diversity but play a role in water quality improvement. Floodplain lacks heterogeneity, and has been landscaped purely for recreational and flood control purposes. The creation of minor depressions along the flood channel would have allowed the creation of seasonal floodplain wetlands Photo credit: N. Newman) (See also colour plate 10)

constricted, stepped gabions were used on the banks instead of the flattened earth base. The first 2 to 3 m of the flattened base were underlain by a lining of high-density, porous polyethylene mesh, laid some 150 mm beneath the soil surface, to safeguard against erosion. Local unskilled labour was used in the construction phase of the project. This was important in the context of high unemployment in the local community.

9.2.4 Project results

The study area was independently assessed in 2002, some two years after completion, and again in 2005. By 2002, *Cyperus textilis* and *Typha capensis* had established themselves in a solid band along the stream margins, together with grasses, *Cotula* sp. *Phragmites australis* reeds and weedy species, such as *Rumex* sp. and *Polygonum* sp. The low-flow channel edging of boulders effectively restricted invasion of the high-flow channel from the water by reeds. During periods of elevated base flows, the low-flow channel provided relatively good quality in-stream riffle habitat, albeit a habitat not naturally found in this system. The entire width of the channel formed wide riffle or stony run habitat, with fringing marginal vegetation providing areas of marginal vegetation habitat. Over time, it is expected that the cobble bed will slowly give rise to sedimented reed beds – this was becoming evident by 2005 (see Figure 9.3). With sediment, however, comes the opportunity for the river to braid and cut shallow channels, thus adding backwaters and sand bars to the range of available habitats. In the early summer of 2005, fifteen different aquatic macro-invertebrate families were found in the river, compared with only five in sections of the river upstream of the remediation area. Although these were all hardy, pollution-tolerant species, and included many air-breathing coleopterans and hemipterans, they nevertheless suggest that in-stream habitat diversity has improved (Day et al., 2005).

By contrast, the gabion structures themselves remained sterile structures with a tendency to collect wind- and water-carried litter, accommodating only occasional weedy plant species. Pedestrian traffic over the gabions has negatively affected the success rate of plantings near the gabions. Although in terms of riverine habitat the gabion weirs have little value, they probably play some role in improving water quality by water aeration. Moreover, given the constraints on channel width, their role in dissipating energy and thus allowing for the soft engineering approach adopted here, as opposed to a hard canal, is also valuable, and their presence should thus be seen in the broader context of controlled erosion, habitat creation and limited spatial availability.

While the in-stream vegetated habitat provides ecologically useful habitat diversity, the planted river edge results in little more than an edging of the river course, between 1 and 1.5 m wide. While this fringe will enhance marginal habitats, providing shelter for riverine animals such as frogs and insects, it will do little in the way of creating a riparian corridor along the river. It is also too limited to buffer the river from impacts like the stormwater drains running down the channel sides – Emmons and Olivier (2001) show in their review of the literature that a 15 to 31 m buffer width is required for effective protection of water quality in a water body. The main reason underlying the lack of attention to creation of floodplain habitat was that, after flood control, the key objective of landscaping the river was to increase amenity value by allowing local communities easy and safe access down to the river margins.

The high-flow channel has proven to be of immense amenity value, with children playing on the grassed base, adults walking along the river bank and families picnicking on the upper bank, looking down onto the river and high-flow channel base. Unfortunately, from an ecological perspective, this was one of the less successful aspects of the project, and one where an opportunity for substantial improvement in the habitat quality of the river has been missed. Seasonal floodplain habitat has all but been lost from urban rivers in Cape Town (Davies and Luger, 1994) and the Moddergat River project provided an opportunity to attempt rehabilitation of this critical wetland habitat type.

In retrospect, some compromise between the needs of the community for recreational space and the spatial requirements for substantial improvement of riverine habitat might have allowed a widening of the planted fringe along the river edge and the creation of nodes of wider wetland and riparian habitats on the channel base, leaving space in between for recreational areas. The wetland nodes could have coincided with stormwater outlets, thus also buffering the river water from potential pollutants, and an area for limited absorption of nutrients and other contaminants in the stormwater. Incorporation of these measures into the project would have had the important function of providing a longitudinal, sheltered riparian corridor, for the movement of small animals through the hostile urban area between the more rural Eerste River downstream and the agricultural area upstream.

Lining the high-flow channel with boulders appears to be an effective means of controlling the invasion of reeds from the low-flow to the high-flow channel. From an amenity perspective, this is an important point, since local communities were opposed to the creation of a dense reedbed channel. Alternative low-growing wetland species (e.g., *Carex* spp.) would have provided water quality, and even habitat functions, without threatening human security.

9.2.5 Stakeholders and their roles in the project

An important aspect of the project design was to involve the local community in the project, particularly in the long-term maintenance of the river corridor. Local schools were thus involved in artwork, tree planting and other educational aspects of the project, and a plant identification guide was drawn up for identifying river plants. The Macassar community was also engaged in discussions on the need to control litter – an ongoing source of litter into the river was, and is, windblown litter from adjacent housing areas, where litter accumulates from overflowing rubbish skips.

Long-term maintenance of the project, which devolves to the local municipality, has been identified as a critical issue. Ongoing clearing of litter and alien vegetation, including kikuyu grass, is required, as well as control of reeds in the low-flow channel, to prevent encroachment by *T. capensis*. Annual clearing will be necessary in time, given that past river maintenance practices, such as dredging of the low-flow channel, will no longer be possible, given the design of the channel.

Roles of the stakeholders in the project implementation were as follows:

- The city of Cape Town (land owner, flood control and safety and security concerns)
- Geustyn Loubser Streicher Inc. (design engineers)
- CCA Environmental, CNdeV landscape architects (public process, overseeing of landscaping implementation, environmental monitoring).

9.2.6 Conclusion and recommendations

The project highlighted the common case of social and ecological objectives in river management being at variance with each other (Day et al., 2005), despite the fact that the engineering objectives of the project were all met. From a social and amenity perspective, the modified river channel provided an aesthetically pleasing, safer environment for recreation than the previous deep channel. Control of dense reed growths has been achieved to a large extent by lining the channel with riprap, which has had the unexpected, at least short- to medium-term, side-effect of limiting lateral expansion of the reedbed. The resulting grassed floodplain is of value in terms of amenity, providing a safe recreational area protected from traffic and far from deep water areas once associated with the channel.

As designed, the river channel did not however address issues like water quality in inflowing stormwater to any real degree. Using parts of the floodplain area for the creation of filtering wetland nodes would have been a useful addition to the project, in an urban scenario where water quality is possibly one of the most pressing constraints to ecosystem function and structure in aquatic habitats. Over time, and with poor implementation of any reed management plan, the low-flow channel has slowly been choked by reed growth, resulting in a dense reeded channel (see Figure 9.4). This type of habitat offers a less diverse habitat to insects and other riverine fauna. It does, however, play a role in water quality amelioration – an aspect that may be of more importance in the context of urban rivers than the provision of diverse habitat, the colonization of which may be limited by water quality.

Landscaping and planting the grassed high-flow channel and upper banks resulted in an aesthetically pleasing riverine amenity used extensively by local residents. From an ecological perspective, however, it is sterile, affording no wetland habitat. Such design objectives would be in direct opposition to the use of the floodplain area for recreation, where safe and secure open spaces for recreation are lacking. In different socio-economic circumstances, the creation of a wide, floodplain area, planted with a variety of floodplain wetland species, permitting walkways but not large playing fields, would have had more appeal. Day and Ractliffe (2002) provide several examples of these.

This highlights the critical importance of effectively marketing ecological design criteria among local communities if opportunities to improve aquatic habitat are to be maximized within the broader context of urban renewal and upliftment.

The theft of plants and vandalism are all ongoing in the river corridor, highlighting the fact that sectors of the community did not take on personal and community ownership of the project. Attempts were made to address such issues in the project implementation phase, which included efforts to educate local children about the river project, with art competitions and tree planting days. However, such programmes need to be extensive and should be integrated with social development and capacity building to engender a more widespread feeling of ownership of the public open space.

A practical management plan for the operational phase of the system will be a necessary long-term requirement. It will need to address issues such as the long-term manual removal of sediment from the river channel, control of invasive reeds and alien aquatic, wetland and terrestrial plant species, and litter collection, the management of which is presently inadequate and results in the accumulation of piles of litter on the roadside that detract from the landscape level enhancement from channel modification and planting.

9.3 REHABILITATION OF THE WASIT NATURE RESERVE, SHARJAH, UNITED ARAB EMIRATES

Elizabeth DAY

University of Cape Town, Freshwater Research Unit, Rondebosch, 7700; South Africa

The Wasit Nature Reserve comprises a broad wetland expanse, located on the outskirts of Sharjah City, which, with some 750,000 inhabitants, has a high growth rate. It is located in a desert region (rainfall *ca* 120 mm/year). The Ecosystem is characterized by remnant saline and hypersaline salt pan and salt marsh mosaic (see Figure 9.5). Threats to the system included large-scale chemical dumping; organic and other wastes; and the perennial inundation and drowning of seasonal wetland habitats. The objective of the project was to rehabilitate the aquatic and terrestrial ecosystems, with the aim of improving tourism, recreational and educational values. Major actions included the removal of dumped waste and berms and the improvement of aquatic habitats by reconnecting flow corridors.

9.3.1 Background

The Wasit wetlands, previously known as Ramtha Lagoon, are located on the outskirts of Sharjah City, in the Sharjah Emirate, United Arab Emirates (AUE), on the Arabian Gulf. They are the remnants of once-extensive coastal salt flats (*sabhka*) and saltmarshes that occurred in areas where groundwater daylights near the coast or at the outlets of ephemeral streams and longitudinal wetland systems. Extensive infilling has taken place in the Sharjah area, resulting in the complete destruction of natural

Figure 9.5 Wasit Nature Reserve (See also colour plate 11)

Photo credit: E. Day

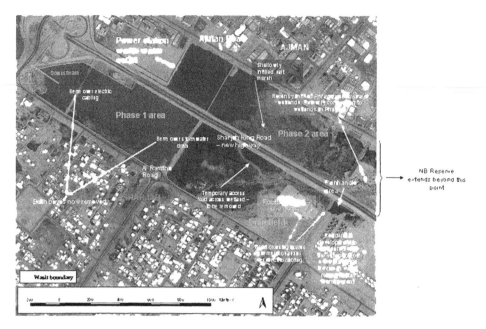

Figure 9.6 Aerial view of Wasit Nature Reserve site, in Sharjah, UAE, bordering Ajman Emirate (See also colour plate 12)

Source: after Day and Ewart-Smith 2005.
Note: Phase 2 border extends beyond right side of the area shown.

flow corridors to the sea downstream of Wasit, and the delineation of the present reserve area by roads to the south, north and northwest (see Figure 9.6).

The most recent intrusion into the wetland as a whole was its dissection, between 2003 and 2005, by the unmitigated construction of the Sharjah Ring Road across the wetlands, without provision for the passage of surface or subsurface flows across the road and between wetland areas (see Figure 9.6). The road essentially split the wetland into two separate portions (referred to as Phases 1 and 2), both of which are closed systems, without any connectivity to downstream reaches, and which back up against the downstream road (Phase 1) and the infilled platform of a power station (Phase 2).

9.3.2 Key aquatic habitat issues in urban water management

In June 2005, the wetlands of the Wasit Nature Reserve comprised a southeast to north-westerly running swathe of highly impacted saline and hypersaline wetlands, with wet-land habitats in Phase 1 ranging from brackish seeps to a mosaic of seasonally saturated and artificially inundated salt flats and saltmarsh ponds, surrounded by dense halo-phytic scrub vegetation, all the way to large, permanently inundated, artificial water bodies. Artificial inundation of saltmarsh areas, caused by backing up water against various berms, resulted in the effective drowning out of saltmarsh vegetation and its replacement by deep, hypersaline pools (Figure 9.5).

Phase 2 was comprised of a broad expanse of mixed terrestrial and wetland habitat, with seasonal to ephemeral wetlands occurring on low-lying areas of the site. The northwestern portion of the site was dominated by permanently inundated pools, while the southeastern section was largely terrestrial and heavily infested with alien mesquite, *Prosopis juliflora* (Day and Ewart-Smith, 2005). Water entered the wetlands mainly as groundwater flows from the northeast, as well as through the dunes from the east (GEOSS, 2006). Satellite imagery and aerial photographs of the site show clear linkages in surface water channels, suggesting that they used to braid across the site. These inflows resulted, in April 2006, in clear pathways of visible trickle flow out of the salt marshes and into the downstream open water bodies.

Prior to the implementation of the rehabilitation project, the following wetland habitat types were identified on the wetland reserve as a whole:

- *Juncus* marsh habitat, in areas where groundwater flows are less saline (< 1000 mS/m)
- *Phragmites australis* reedbed wetlands in nutrient enriched and brackish conditions (electrical conductivity up to 3000 mS/m)
- Shallow seasonally inundated pans
- Hypersaline marsh ponds and impacted salt marsh mosaic
- Shallow hyper-saline seepage channels
- Infilled salt marsh habitat
- Large, permanent open water bodies (Day, 2006).

Water chemistry data showed that the artificially inundated open water ponds in the lower reaches of the wetlands were hypersaline and, without intervention, were on a trajectory of increasing salinity. The inundation of the remaining saltmarsh mosaic upstream in Phase 1 had occurred as a result of more recent impoundment against downstream berms. This exposed these habitats to the poor water quality in the open water systems downstream and resulted in large-scale die-off of permanently inundated saltmarsh vegetation.

Nevertheless, the wetlands, even in their impacted form, were assessed as having high conservation value, partly as an example of arid saltmarsh systems – systems that have been described as 'undervalued natural assets and, significantly, under-represented in formally conserved wetland areas, such as listed RAMSAR wetland sites' (Williams, 2002). Their key importance lay in their provision of habitat for a diversity of migrant, over-wintering and resident birds (Aspinall, 1998). From a social perspective, the wetland area also had high potential, as a result of its proximity to a large urban centre, the variety of natural habitats available in the reserve area, and the iconic value of the Greater Flamingo, a bird once found there in large numbers, which could in theory be encouraged to return to the site in similarly large numbers.

9.3.3 Objectives of the case study

Following the basic assessment, which highlighted the present and potential importance of the Wasit wetlands for conservation, rehabilitation measures for the Wasit wetlands were formulated with the objectives of ensuring sustainability of the present wetland system and enhancing its aesthetic and recreational appeal.

A condition for all recommendations in the rehabilitation plan were that the envisaged future nature reserve at Wasit had to be sustainable and should not have deleterious effects on existing ecosystems that were conservation-worthy in their own right (de Soyza et al., 2002). This meant that management and rehabilitation activities at

Wasit would need to take place within a framework that focused on the conservation and, as far as possible, restoration of natural habitat types, facilitating artificial habitats for flagship species (such as Flamingos) taking secondary importance.

9.3.4 Implementation of the project and involved stakeholders

The major drivers of wetland function and habitat quality at Wasit were identified as salinity and inundation (Day and Ewart-Smith, 2005). Of these, salinity and the associated loss of freshwater inflows into the system were both considered critical to the sustainability of the wetlands as aquatic habitats, and rehabilitation measures centred on addressing these issues. Additional measures were also recommended, with the aim of improving habitat quality and diversity, by addressing seasonal flow and water quality regimes within Wasit, and by expanding the conserved area to include upstream corridors and important natural habitats. The following major intervention measures were recommended:

- Installation of a drainage structure in the lower wetland to allow initial drainage of polluted, hypersaline water from the effectively closed-off wetland system, in order to prevent the ongoing accumulation of pollutants including dissolved salts in the wetlands and to allow the regulation of maximum water levels in the wetland as a whole, thereby facilitating the rehabilitation of drowned saltmarsh habitat and increasing areas of shallow wading habitat upstream.
- Reconnection of the Wasit wetlands with source areas of fresher water upstream as follows:
  - removal of dumped material in the former reedbed channel upstream of the new highway
  - removal of impediments to surface flow in Phase 2 including large berms and infilled channels at road crossings
  - insertion of a system of pipes beneath the highway, to allow passage of surface and groundwater flows from upstream source areas, identified on aerial photography (see Figure 9.7).

The following measures were considered essential for the long-term sustainability of the wetlands – failure to implement them is likely to result in long-term shrinkage of wetland habitats downstream and ongoing degradation in habitat and water quality:

- Connection of the impounded open water pond north of the highway, in Phase 2, to wetlands in the lower reaches of Phase 1, to permit drainage and improve water quality
- Protection of sources of fresh ground- and surface water in the vicinity of the site
- Removal of dumped material from the site, with priority being given to chemical or organic material
- Re-landscaping of disturbed areas, affected by rubble removal and compression
- Removal of berms resulting in the inundation of saltmarsh habitats upstream – primarily the electric cabling berm, the stormwater berm and the temporary access road, used during the construction of the new highway
- Use of berm material to create islands in the large, impacted open water areas in the downstream extent of the wetland where habitat diversity is low and water is deep for wading birds

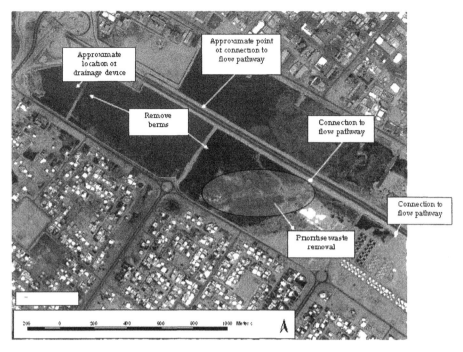

Figure 9.7 Action plan for the rehabilitation of the Wasit wetlands (See also colour plate 13)

Source: after Day and Ewart-Smith 2005.

● Fencing off the reserve, as a measure to control dumping and, importantly, allow control over domestic and feral predators
● Removal of alien mesquite (*Prosopis juliflora*) from the site and from drainage corridors on Phase 2 in particular, since this species abstracts groundwater that would otherwise be available to wetlands downstream.

Once the basic rehabilitation of the wetland had occurred and its protection was ensured through fencing and the establishment of a management regime, additional measures were recommended, some aimed at long-term management and others at the creation of additional habitat types and aesthetic improvements.

There were also measures to protect the reserve from the impacts of surrounding land-use which included:

● noise pollution associated with adjacent highways
● polluted runoff from road surfaces
● seepage from septic tanks and sewerage systems (both on- and off-site)
● intrusion by feral and domestic cats, dogs and other predators
● invasion, grazing and nutrient enrichment by goats and other livestock
● abstraction of surface- and groundwater from relatively fresh source areas higher up in the catchment.

Involved stakeholders included His Highness Dr. Sheikh Sultan bin Mohammed Al Qasimi, Ruler of Sharjah (project initiator and funder), Gary Bartsch International (project conception and co-ordination) and various biophysical specialists, including the Freshwater Consulting Group (wetland ecosystems), GEOSS (geohydrology) and Drs. Hellyer, Howorth, Aspinall, Conrad, Day and Ewart-Smith.

9.3.5 Project results

The proclamation of the Wasit Nature Reserve and its subsequent fencing off has probably had a negative effect on local communities, some of whom used the area for grazing sheep and worship (on the dune areas). It has, however, generated much positive interest in the wider area of Sharjah and adjacent emirates, where it has the prestigious position of being the only wetland reserve in Sharjah. In an area where there had until recently been little effort to develop an understanding of and appreciation for the natural environment, the availability of environmental educational resources in and around Sharjah is limited. In this context, the proclaimed reserve has the potential to provide a wealth of educational opportunities to local schools and families, as well as providing an alternative outdoor entertainment niche for tourism. Naturally, there is a degree of conflict between the type of facilities required to attract visitors to the reserve area and the ecological perspective of conservation and rehabilitation of important habitats. Since the success (and thus the ongoing funding for management and ongoing rehabilitation of the reserve) is measured to a large degree by the number of visitors that will use it, creating an attractive facility assumes a high level of importance. Moreover, given the harsh climate of the area, a large proportion of the visitor's experience of the reserve needs to take place under cover, preferably equipped with air conditioning. This means that the visitor centre will have a sizeable ecological footprint. It will also have its own requirements for freshwater, management of sewage waste, provision of space for parking – all within the confines of a reserve area.

Other areas of potential conflict lay in the degree to which the reserve should be based on the conservation and rehabilitation of the natural wetland and dune ecosystems, or to what extent it should be used as an area in which a range of other systems could be developed and presented as a kind of ecological theme park. Whilst this approach would have definite tourist appeal, it would also largely defeat the objectives of the rehabilitation programme – and might have occurred more cheaply elsewhere, without the need to remove hundreds of tonnes of rubble. The decision was taken to focus on the rehabilitation of the natural ecosystems.

The other main areas of conflict between ecological and social requirements revolved around local perceptions that natural attractions need to look bright, green and colourful – features that are in opposition to natural desert wetland habitats. The provision of a fast-growing, green, attractive screen around the reserve area, with an additional dual purpose of providing a noise barrier from the highway, resulted in the selection of a variety of plants, some of which were non-native and had low tolerance to highly saline conditions. Fast growth was secured, after long debate, by providing low-salinity irrigation water to plants along the wetland perimeter. Irrigation water was sourced from treated sewage effluent water, making ongoing nutrient enrichment of the system a threat.

This case study has been compiled only one year after the initiation of the rehabilitation process, and comments on changes to date must be seen in the context of a short-term response to long-term changes. This said, the implementation of rehabilitation activities in the Wasit Nature Reserve has had profound implications for the structure and functioning of Phase 1 wetlands. Positive effects that have occurred so far include the following:

- A pronounced visual improvement has been remarked in Phase 1 habitats as a result of large-scale removal of waste;
- The clearing of waste has also removed a large proportion of sources of chemical pollution on the site – a reduction in nitrogen-based pollution was observed in 2006 chemical data.
- The lowering of the hypersaline water table in the southern portion of Phase 1, as a result of the removal of the stormwater and electric cable berm across the open water wetland has had a pronounced effect on water quality, wetland habitat and wetland community composition. Changes in water chemistry and inundation have resulted in the creation of nodes of fresher water habitat within the saltmarsh area. These habitats have the highest diversity of invertebrate species on the site as a whole. In 2006, six months after the start of the rehabilitation programme, there was a noticeable increase in bird activity on the wetland; three new species were observed on site for the first time; and the Western Reef Heron bred on site – also a first for Wasit (Tovey and Aspinall, 2006).

To date, rehabilitation activities have thus led to a marked improvement in wetland function. However, if the wetland area is to be sustainable in the long term, fundamental aspects of the recommended rehabilitation programme still need to be implemented:

- Provision of a controllable outlet structure to prevent the impoundment of water and prevent the long-term build-up of contaminants
- Provision of pipelines beneath the new highway to reinstate connectivity with groundwater sources and sporadic surface flows from Phase 2 into Phase 1
- Long-term monitoring to inform a management system with regard to setting upper water levels and provision of seasonal variability.

9.3.6 Conclusion and recommendations

At the same time that the positive impacts of the rehabilitation project are realized, it is also noted that the cost implications of rehabilitation at such a scale, and from such a degraded starting point, are enormous. Moreover, certain fundamental issues still need to be addressed, again at no small cost. Passing pipes beneath the highway to allow the passage of subsurface flows, for example, is expensive – but unless this and the other important activities identified as high priorities are actually implemented, the costs borne so far in wetland rehabilitation will be largely cosmetic, and the long-term trajectory of the wetland will remain one of ongoing degradation.

The project also highlights some of the practical issues that arise when ecosystem rehabilitation is funded with the objectives of simultaneously creating public amenity and recreational facilities. Compromises in the final product need to be made, and a lot of these revolve around timescales. Ideally, ecological rehabilitation should take place in phases, allowing first subtle manipulation of the system drivers followed by periods of monitoring to assess the impact of intervention on the system. The luxury of a phased long-term implementation project is not, however, available when the success of the product is largely being judged on the basis of its appeal to visitors – and the compromises entailed in fast-tracking the opening of the system (e.g., the use of low salinity irrigation water; import of non-indigenous plant stock; sourcing of plants from different genetic stock; etc.) risk compromising the final ecological outcome. By the same token, however, if funding is withdrawn because major funders lose interest in a project that is not showing visible results for a considerable financial investment, the ecological outcome will also be affected.

9.4 THE ECOHYDROLOGICAL DIMENSION OF SMALL URBAN RIVER MANAGEMENT FOR STORMWATER RETENTION AND POLLUTION LOADS MITIGATION: LODZ, POLAND

Iwona WAGNER[1,2], Jan BOCIAN[3], Maciej ZALEWSKI[1,2]

[1] European Regional Centre for Ecohydrology under the auspices of UNESCO, Polish Academy of Sciences, 3 Tylna Str, 90-364 Lodz, Poland
[2] Department of Applied Ecology University of Lodz, 12/16 Banacha Str, 90-237 Lodz, Poland
[3] ECOINNOVATIOS, consulting company, Lodz, Poland

This case study addresses small urban streams (average flow $< 1\,m^3s^{-1}$) integrated with the city sewerage in Lodz, Poland. Their channelization reduces self-purification capacities and deteriorates water quality. Highly impermeable catchments reduce water retention in landscape and increase stormwater peak flows. This periodically impacts the purification capacity of the wastewater treatment plant. The case study's objectives include the application of ecohydrology in order to address the above issues. Their harmonization with technological measures aims to improve residents' health, quality of life and economic benefits.

Figure 9.8 Impoundment on the Sokolowka River (See also colour plate 14)

Photo credit: W. Fratczak.

9.4.1 Background

Environmental hydrological setting

Urban water management in Lodz is challenged by its specific hydrological situation determined by geographical setting. Although located in the plain district of the country, the city itself lays in a relatively steep area (ranging between 180 to 235 m above sea level within the city) and is divided into 18 catchments drained by streams with average flows of $< 1 \, m^3 s^{-1}$ (see Figure 9.8 and Figure 9.9). Rapid expansion of the city in the first quarter of the nineteenth century contributed to severe degradation of natural resources: the streams were channelized and covered. They were adapted as sewerage channels and continue to play this role. The slope of the rivers and channels, up to 5 to 7 ‰, contributes to accelerated runoff and high flows during wet weather.

General characteristics of the urban area

With 776,000 inhabitants as of 2004, Lodz is the second biggest city in Poland. Recent developments in education, scientific and high-technology centres, have changed the city's profile and its inhabitants' expectations, introducing the anticipation of a higher quality of life, including the development of infrastructure, the revitalization of industrial architecture and environment rehabilitation.

The total area of the City covers $294 \, km^2$ and is characterized by a regular street pattern (Klysik and Fortuniak, 1999). The central part of about $15 \, km^2$ contains the oldest development from the period of the textile industry boom and is more than 80% impervious. The largest part of the city, surrounding the centre, houses a combined residential and industrial development, with a density of 20,000 inhabitants per km^2. The outskirts are occupied by districts with detached houses and much lower imperviousness (less than 25%). Many of these districts are adjacent to meadows, forests (7% of the total area) and arable land (Statistical Yearbook for the Lodz Province, 2005).

Brief Urban Aquatic Habitat (UAH) inventory and prioritization

The contrast between the out-of-town zone and the built up area, versus that of less developed suburban sections with residential buildings, sets a template of potentials for and constraints on the ecological restoration of aquatic habitats. Semi-natural sections of rivers in suburbs possess a greater capability for successful restoration and may provide examples for further up-scaling of the tested ecohydrological measures. The densely developed centre of the city imposes more constraints. Some of the rivers cannot be restored any longer, as they have disappeared due to regulation, separation from flow sources or a high rate of impermeable surfaces in their catchments. Others are covered and flow under historical or strategically important buildings. Restoration measures in this zone must strongly rely on technical solutions and limited possibilities of sewerage reconstruction, which can happen only in combination with an integrated programme of environmental and architectonic revitalization.

Urban drainage and wastewater management

Lodz is serviced by a mix of two sewer systems – combined sewers in the centre and separate sewers in the suburbs (Zawilski, 2001). The industrial and domestic sewage

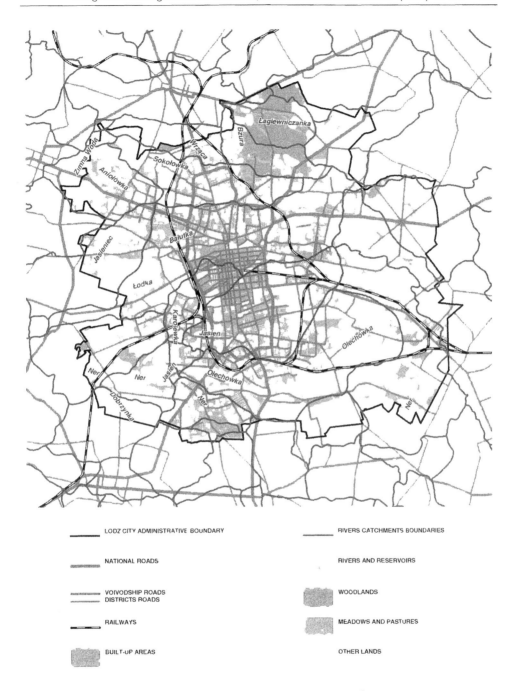

▬▬▬ LODZ CITY ADMINISTRATIVE BOUNDARY	▬▬▬ RIVERS CATCHMENTS BOUNDARIES
▬▬▬ NATIONAL ROADS	RIVERS AND RESERVOIRS
▬▬▬ VOIVODSHIP ROADS / DISTRICTS ROADS	WOODLANDS
▬▬▬ RAILWAYS	MEADOWS AND PASTURES
BUILT-UP AREAS	OTHER LANDS

Figure 9.9 Streams catchments in the territory of the Lodz City (294 km^2) – the Sokolowka (northern part of the city) and Ner rivers (southern part of the city) serve as demonstration areas for implementation of ecohydrology in systemic urban water management (elaborated by: M. Grzelak based on the resources of the Marshal`s Office of the Lodz Voivodship) (See also colour plate 15)

(from about 80% of the population) as well as stormwater (in combined sewers) are conveyed to the central city Wastewater Treatment Plant (WWTP). After physical and biological treatment, treated sewage (average flow $2.5 \, m^3/s$) is discharged into the Ner River, with a natural discharge of less than $0.8 \, m^3/s$. The flow of sewage increases more than ten-fold during storms, exceeds the treatment capacity of the WWTP and impacts on the quality and ecological stability of the river. The wastewater treatment plant produces 70,000 tonnes of sewage sludge per year (200 tons/day), which is digested and stored in lagoons. The forming biogas is de-sulphurized and used to cover some energy requirements of the WWTP operation.

Institutional setting and projects related to aquatic habitats management

The key actor in water management and environmental protection in Lodz is the City of Lodz Office. An important role is played by the Municipal Company of Water Supply and Sewage Systems, which produces and provides drinking water to the city and delivers sewage to the WWTP. The city's 2,000 km of water-pipe network and 1,600 km of sewers are maintained by the Lodz Infrastructure Company. Both companies are supervised by the City Council, which establishes the framework for their functioning and makes strategic decisions.

There is a long tradition of incorporating ecological measures into integrated urban water management (IUWM) in Lodz. Some impoundments were created in natural sections of rivers within parks and in suburbs, working both as stormwater retention and recreational areas. One of the major activities has been a 'Small Impoundment' project of the City of Lodz Office, operating since 1997. This activity started valid discussions on the importance of integrating urban water management with aquatic habitats restoration and Adaptive Environmental Assessment and Management (Holling, 1995) in the future development of the City. Since 2000, complex analyses have been conducted as General Projects for River and Stormwater management, developed for each river.

9.4.2 Key aquatic habitat issues in urban water management

Adaptation of small city rivers for interception of stormwater loads

The city's location, catchment morphology and development reduce water retentiveness in the landscape. The effects of these activities are especially pronounced in small urban catchments, where streams and rivers are channelized for stormwater detention according to traditional engineering rules. As listed by Bocian and Zawilski (2005), disturbances of the water cycle in Lodz include the acceleration of surface flow and increase of flood peaks; a decrease of groundwater recharge; and decrease of extent and duration time of low flows.

Degradation of rivers habitats and of their self-purification potential

The degradation of freshwater habitats by their channelization seriously handicaps ecosystem functioning by impacting their biotic structure, biological diversity and destroying functional links within the ecosystem. Those impacts, superimposed by the disturbances listed in the previous section, reduce ecosystem resilience, impacting self-regulatory processes. Reduced self-purification, together with high pollutant loads, lowers the assimilative capacity of streams and deteriorates water quality. As a result, toxic cyanobacterial blooms were observed in some reservoirs (Izydorczyk, unpublished data), and the concentrations of the produced toxins reached extremely high values of up to 60 ug/l in 2006 (Jurczak, unpublished data).

Decreased ability of the WWTP for combined sewage purification and sewage sludge utilisation

Increased inflow into the WWTP during flood peaks reduces plant's ability to purify sewage and lower the water quality in the receiving water, the Ner River. Years of sewage disposal have contaminated river and floodplain sediments with heavy metals and organic compounds. The total heavy metal concentrations in the floodplain soils range from: 11 to 598 mg/kg for Zn, 7 to 390 mg/kg for Cr, 7 to 121 mg/kg for Cu and 11 to 93 mg/kg for Pb (Bocian, unpublished data), forming a positive correlation with the organic matter content. Decreases in water use and sewage disposal within the last 15 years have lowered groundwater level and accelerated sediments mineralization, resulting in the leaching of heavy metals and their accumulation in food chains in floodplains used for agriculture. This process is enhanced by high combined sewerage flows through the WWTP in wet weather. The use of sewage sludge produced by the WWTP constitutes another economic and ecological issue.

9.4.3 Objectives of the case study

The objective of this project is elaboration and validation of the ecohydrology approach (Zalewski et al., 1997) in urban water management. According to Zalewski (2006), implementation of the concept is based on the three fundamental tenets: using synergies between the catchment water cycle and dynamics of its biotic component; harmonizing existing and planned hydrotechnical solutions with ecological biotechnologies; and integrating complementary synergistic measures at all scales.

By extending the use of technical measures in urban water management with ecological solutions, both the quality of environment and cost-efficiency of management practices can be improved, and some additional income can even be generated. To assure the project's sustainability, appropriate attention has to be also given to social aspects, enhancing understanding, appreciation, acceptance and active participation of the public and decision-makers.

The long-term objectives (more than five years) of the project include the following:

(i) *Environmental goals*
   – rehabilitation of selected sections of the city's rivers based on harmonization of hydrotechnical and ecological measures for increasing hydrological and ecological capacity of urban aquatic ecosystems in coping with human impacts
   – reclamation of degraded urban river floodplains and wetlands and natural measures for stormwater retention
   – increased number of green areas and open waters in the city for improvement of microclimate, human health and quality of life
   – providing ecological solutions complementary to operating technologies in the Wastewater Treatment Plant in Lodz, increasing the ecological and economic

efficiency, according to EU requirements (using sewage sludge, bioenergy production, protective green zone around the WWTP and capital savings).

(ii) *Societal goals*
- providing solutions for secure water resources for social safety (preventing toxic algae blooms, heavy metals contamination)
- providing ecological solutions complementary to operating technologies in the WWTP for increasing ecological and economic efficiency and meeting the EU directives requirements (e.g., using sewage sludge, bioenergy production, protective green zone around the WWTP)
- ensuring strategy for the development of high quality recreation areas in a good state environment
- increasing awareness of good quality water and environmental resources value and their role for the quality of life
- attracting new investments and people of action, contributing to increasing city attractiveness, creating strong city image and local civism
- improving internal (within the city stakeholders) and external (national and international) cooperation, enhancing knowledge sharing, experience flow and broad city promotion
- building a basis for participatory approach to environmental and city problems, promotion/strengthening of bottom up initiatives.

(iii) *Policy and Administration goals*
- defining clear priorities in water management and related sectors and accelerating investments in the related infrastructure
- incorporating IUWM know-how and ecohydrological innovations into local (city-level) legislation and policy and then up-scaling to the national level
- adopting water management and related issues to EU standards and requirements
- improving efficient cooperation and information sharing between stakeholders, governmental and self-governmental authorities and services on the city water resources management strategy
- developing strategies for fundraising.

9.4.4 Project implementation and results

The implementation of the project covers the three following aspects:

(i) adaptation of small city rivers and catchments for interception of stormwater and pollution loads
(ii) elaboration of comprehensive concept of WWTP management addressing issues of sewage sludge utilization for enhancement of biomass and bioenergy production, economic feedback and river rehabilitation
(iii) providing socio-economic feedback to city inhabitants based on the use of ecosystem resources of regenerated urban ecosystems, improvement of health and quality of life (Zalewski and Wagner, 2006).

The selection of two pilot catchments (Table 9.1) allowed tackling a comprehensive scope of water-related issues specific for the city and validate the concept.

Table 9.1 Catchment characteristics of the two demonstration project rivers in Lodz

Feature	Sokolowka	Ner
Catchment area [km²]	44,5	1866,5
River length [km]	13,4	129,5
River channel regulation [%]	100	75
Impoundments on the main channel	4	–
River channel gradient [%]	0,55	0,03–0,2
Annual rainfall [mm]	535	499
Annual discharge [mm]	153	168
Flow regime [m³/s] average (min – max)	0,17 (0,03–15)	9,98 (0,74–97)

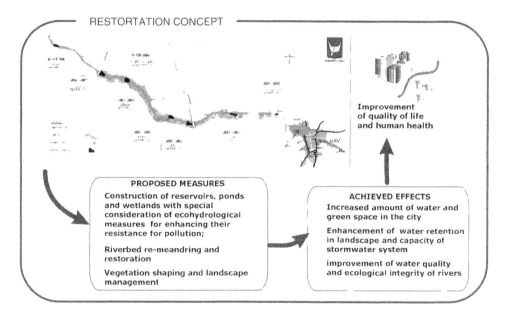

Figure 9.10 The general concept for municipal river restoration, including the ecohydrology measures implemented in Lodz

Source: Zalewski and Wagner. 2006. based on the material of the City of Lodz Office.

Project 1: Restoration of a municipal river for stormwater management, increase of water retentiveness and improvement of quality of life

The Sokolowka River crosses the northern part of the city. The river's natural flow had gradually disappeared, replaced nowadays mostly by discharges from stormwater outlets. The channel was regulated by concrete slabs in order to straighten the course and deepen the bed for runoff detention. Nevertheless, the middle section of the valley maintained semi-natural patches of meadows, wetlands and forests and that made this section appropriate as a pilot area for application of ecohydrological measures (Figure 9.10).

Increasing stormwater retention: several measures were proposed for adapting the rivers for stormwater and pollution loads interception (Bocian and Zawilski, 2005).

Those related to ecohydrology include, among others, reconstruction of semi-natural channels, construction of on-stream ponds, application of phytotechnology for controlling hydrological regime and water quality, creation of local wetlands for annual flow control, improving conditions for river rehabilitation and improving water quality at drainage outlets (Bocian and Zawilski, 2005; Zalewski and Wagner, 2006).

Ecohydrological adaptation of hydrotechnical infrastructure: the assimilative capacity of constructed ponds and reservoirs can be enhanced by ecohydrology measures, such as adjusting reservoir hydrodynamics; shaping the biotic structure of reservoirs to increase their capacity against pollution and eutrophication symptoms; and adapting the hydrodynamics of biofilters at stormwater outlets for allocating nutrients into unavailable pools and preventing the flushing of pollutants during storm flows. Water quality is particularly important, since the reservoirs also serve as recreational areas for nearby residents.

Ecological restoration of selected river sections: the self-purification ability of rivers can be increased by restoration of the habitat (see Chapter 5) and its related biotic structure in land-water ecotones, floodplains, wetlands and adjacent landscape. The experiments in Lodz's rivers show that adapting vegetation communities to habitats' hydrological features may increase their ability for providing the required services (see Figure 9.10).

Supporting development of green areas in the city and increasing the quality of life: plans of greenlands and open water area extension (as presented in Chapter 6) will increase water retentiveness, contribute to stabilizing the water cycle and improve microclimate. In a feedback, this situation will further stabilize vegetation in the city, especially during dry weather. Microclimate improvement may decrease the frequency of allergies and asthma cases and contribute to increasing the population health.

Project 2: Sewage system management for environment quality and positive socio-economic feedbacks (bioenergy production)

The efficiency of wastewater treatment plants in pollutant removal, although meeting legislation norms, is very often not sufficient, if the pollutants concentrations exceed assimilative capacity of the receiving aquatic ecosystems. Nevertheless, very costly wastewater treatment plant construction and operation has to be carried out by local communities. This project, developed in the Ner River Valley and the Protective Zone of the WWTP (see Figure 9.12) focuses on two aspects: the reclamation of a river and its floodplain by means of ecohydrology and phytotechnology and the development of an integrated strategy for restoring cycles of matter and pollutants in the city.

Reclaiming a river and its floodplain by application of ecohydrology and phytotechnology:
Willows and native plant communities were tested for their efficiency in preventing the release of toxic substances and heavy metals from the Ner River and its floodplain. Pilot experiments evaluating growth rate and heavy metal uptake and their effect on the groundwater level showed that phytoremediation measures could be optimized by an ecohydrological adjustment of the vegetation to hydrological characteristics (e.g., groundwater level) of the valley. The observed difference in the growth rate and biomass yield in the first year varied twenty times depending on tested varieties (see Figure 9.11.a). This result shows how important it is to implement a careful adaptation of a plant species to local hydrological conditions in order to optimize the produced biomass

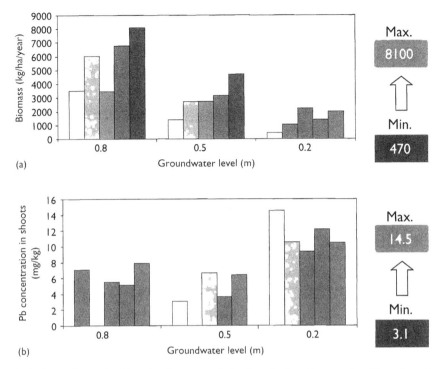

Figure 9.11 Groundwater level as a key factor regulating functioning of energetic willow plantation and floodplain restoration

Source: Bocian, Sumorok, Zawadzka, unpublished.

yield and, consequently, income from bioenergy. In general, the most favourable conditions were achieved at an average groundwater level lower than 0.4 measured from the ground level, while the growth rate at the initial stage depended on shallow and stable water levels. On the other hand, the efficiency of the phytoremediation process depended on heavy metals availability for plants and increased in anoxic conditions of high groundwater levels. Therefore, the remediation potential was determined by a combination of the growth rate and the availability of heavy metals to plants (see Figure 9.11.b). The regulation of groundwater levels or the selection of proper plant communities may be then a tool for optimizing plantation efficiency for either remediation of bioenergy production. Using native plant communities can help in the restoration of biodiversity and ecological values of the river and its valley.

Developing integrated strategy for restoring cycles of matter and pollutants in the city: Sewage sludge, if meeting legislative norms for heavy metals and nitrogen content, can be used on willow plots isolated from the river valley (e.g., in the Protective Zone of the WWTP). Because of high nutrient concentrations, it is a valuable fertilizer for producing biomass, which in turn provides alternative energy (bioenergy) and reduces CO_2 emission, contributing to meeting EU requirements on the promotion of renewable energy sources (Directive 2001/77/EC). The production of bioenergy can lead to

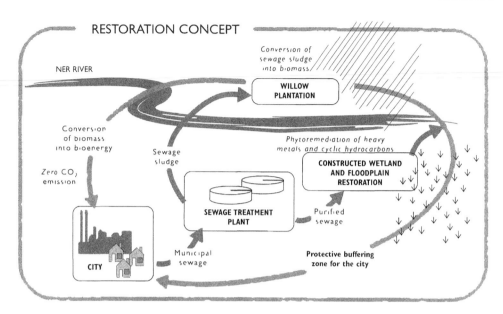

Figure 9.12 Restoration concept for sustainable sewage system management: Phytoremediation for sewage sludge utilization and bioenergy production

Source: Zalewski and Wagner, 2006.

a quicker return on invested capital, the generation of new jobs, the reduction of capital outflow for fossil fuels, and increased revenue for the local economy. The preliminary calculations show that wood chips from short rotation forestry (SRF) of the area of about 3,000 ha could cover the energy needs for municipal buildings in Lodz and eliminate the problem of sewage sludge use (Bocian, 2004).

9.4.5 Stakeholders and their roles in the project

The attempts to solve the water-related issues in the area by elaborating a system approach have been made on the basis of the long-term cooperation between scientists (University of Lodz, International Centre for Ecology – Polish Academy of Sciences, Technical University of Lodz) and the Municipal Office of Lodz. Recently, the activities have been enriched by the European Integrated Project of the Sixth Framework Programme, SWITCH. The Learning Alliance (LA) established within this project assembles the self-government, local national authorities, the city's liaisons, scientific institutions, NGOs, schools, media and society representatives. The goal of the Lodz LA is to identify water-related issues, ensure the efficient implementation of recent scientific results and innovation and ensure their relevance to the local needs. Involving a variety of stakeholders provides functional mechanisms to facilitate implementation of the innovative solutions (Moriarty et al., 2006).

Acknowledgments

The presented concept was elaborated by the Department of Applied Ecology of the University of Lodz and the European Regional Centre for Ecohydrology under the auspices of UNSECO, Polish Academy of Sciences. The Technical University of Lodz and the AQUAProject s.c company contributed crucial efforts to its elaboration. The important step in participatory decision-making was achieved thanks to cooperation with the Sendzimir Foundation. Agrobränsle AB provided willows for experiments. Lodz is a Demonstration City in the Sixth European Union Framework Programme Integrated Project, SWITCH: Sustainable Water Management Improves Tomorrow's Cities' Health (GOCE 018530).

9.5 INTEGRATING ECOLOGICAL AND HYDROLOGICAL ISSUES INTO URBAN PLANNING IN THE ADIGE RIVER FLUVIAL CORRIDOR, ITALY

Maria Giovanna BRAIONI[1], Anna BRAIONI[2], and Gianpaolo SALMOIRAGHI[3]

[1] Department of Biology, Padova University, Via U.Bassi 58/B, 35121 Padova, Italy
[2] Braioni Architect Office, vc. Ponte Nuovo 9, 37126 Verona, Italy
[3] Department of Evolutionary Biology, Bologna University, Via Selmi 2, 40126 Bologna, Italy

About 48% of the population in the Adige Basin (1,637,500 inhabitants) is concentrated in highly urbanized areas. The basin with continental climate in the upper sections, sub-littoral in other parts, contains Natura 2000 sites with rich, endangered biota. River canalization, hydropower production, irrigation, pollution and excavations in the riverbed and riparian areas have resulted in aquatic habitat simplification and threatened biodiversity (see Figure 9.13). The objective of this case study is to balance the above uses while improving landscape quality and bio-assimilation capacity. Major actions include integrating ecological and hydrological issues into urban planning by a procedural/action abacus as a tool for management actions.

Figure 9.13 A stretch of the Adige River fluvial corridor (See also colour plate 16)
Source: Turri and Ruffo, 1992.

9.5.1 Background

Environmental context

The Adige River Basin (11,954 km^2) is located in the Northeast of Italy, in the Alto Adige, Trentino, and Veneto Regions. According to the EU Water Framework Directive (Directive 2000/60/EC) classification, the Adige River Basin corresponds to Ecoregion 4 (Alps) in the northern portion of its mountainous basin and Ecoregion 3 (Italy, Corsica and Malta) in its foothill and plains sections. The altitude typology is high (above 800 m), middle-altitude (between 200 and 800 m) and lowland (below 200 m) (Braioni et al., 2002). The geological composition varies from siliceous in the uplands to calcareous in lower sections. The length of the river is 409 km, and it varies in flow, width, depth, form, slope, shape of the riverbed (natural, embanked, channel, elevated) characteristics and valley shape, mean substratum composition, and adjacent land-uses (Braioni et al., 2002). The model of the integrated management and planning of the Adige River Fluvial Corridor (ARFC) System was tested in eleven sub-sections of the Basin (see Figure 9.14).

Historical perspective

There are many historical references about this region. Austrian cadastre maps supply precise land-uses of the fluvial corridor of the Adige River in the eighteenth and nine-teenth centuries. The ancient riverbed analysis allowed for the reconstruction of his-torical diversions of the Adige in the mountainous fluvial corridor and in the Veneto Plain (Nicolis, 1898). Hydrogeological studies and research highlight how the surface water of the Adige River and the alluvial plain of the basin support groundwater and deep waters of the Fissero Tartaro Canal Bianco Basin (Dal Pra et al., 1991). Other research was carried out in the same period (Hydrodata and Beta Studio, 1999), together with studies about the integrated qualities of the Adige river.

Numerous floods between the end of the eighteenth century and the beginning of the nineteenth century and continuous demands for cultivated land contributed to the straightening of numerous bends in the mountainous stretch and plains. This resulted in the reduction of the river course by 18 km and the construction – within the final above-terrain-elevated stretch (110 km) – of high banks and elevated floodplains (Miliani, 1937).

In previous centuries, many human-caused and natural problems accumulated along the fluvial corridor of the Adige River. Serious damage has caused continuous deterior-ation of the environmental characteristics in the basin. Point and non-point pollution of surface waters and groundwater, loss of the autochthonous riparian vegetation, excava-tions in the river-bed and on the banks, river canalization, and flow diversions were not compatible with maintaining minimum vital flows, and the mitigation of flood risk, geo-morphological instability and landslides (Braioni and Ruffo, 1986).

Current Urban Water Management

Socio-economic and environmental issues are strictly interdependent with each other and determine the goals for Integrated Urban Water Management, involving authorities representing different levels of government. Integrating these activities is very import-ant, as risks, problems and threats may increase if the management and planning of the

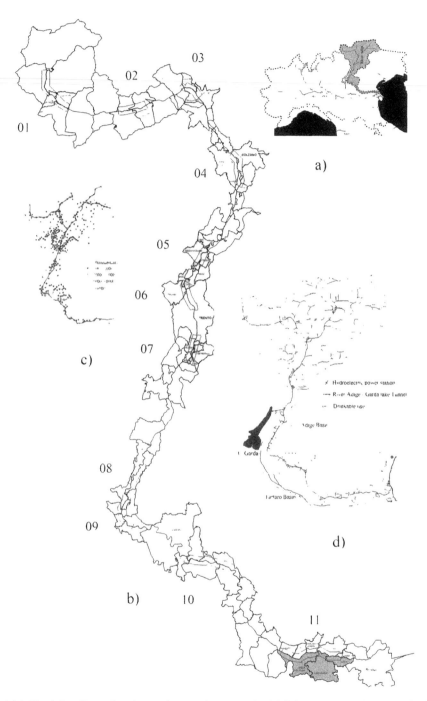

Figure 9.14 The Adige River fluvial corridor: (a) Location (b) 11 Sub-sections analysed in the Basin (c) Concentrations of urban settlements (d) Adige basin, Adige River – Lake Garda tunnel, hydroelectric power station and locations of drinking water withdrawal points

Source: from Braioni et al., 2005a.

whole basin is not comprehensive, because of limited resources and conflicting interests of stakeholders in the basin.

Today, flood risk is defined by the National Authority of the Adige River (Autorità Bacino Adige, 2006). Very high flood peaks (500 m³/s) are diverted into the oligo-trophic Garda Lake by the Mori-Torbole tunnel. According to the law 183/89 the Basin Authority is in charge of pursuing integrated planning interventions at the basin scale, thus finding a common intent, goals for public benefits and preparation of a basin plan with an effective implementation schedule.

The European Community has also developed the Water Framework Directive (WFD) with the purpose of establishing a framework for promoting sustainable water uses, based on a long-term protection of existing water resources, and mitigating the effects of floods and droughts. This directive proposes a protocol with principles and technical-scientific methods for the evaluation of the river ecological status, which will be integrated in the elaboration of the River Basin Management Plan.

This research strives to integrate the evaluation, management and planning of the ARFC system. The Management Plan of the Adige River Basin was financed by the National Basin Authority. Local studies of the Adige River basin were carried out with the funds of the Veneto Region and the Autonomous Provinces of Trento and Bolzano. They were coordinated, integrated and complementary to those conducted by the National River Authority for the Adige river basin.

9.5.2 Key aquatic habitat issues in urban water management

About 48% of the population in the Adige Basin (1,637,500 inhabitants) is concentrated in highly urbanised areas around the cities located along the Adige River fluvial corridor (Figure 9.14). The resident population increases by 500,000 people in winter and summer due to tourism in urban and rural settlements (Autorità Bacino Adige, 2006). Several infrastructure elements such as motorways, railways, hydroelectric and irrigation channels are located along the fluvial corridor. The river basin and its hydro-logical regime are affected by all kinds of hydroelectric, irrigation, industrial, agricultural and recreational, sanitation and water supply uses, with the last one being particularly frequent in the plains (Figure 9.14d) (Turri and Ruffo, 1992). The analysis of macrobenthos, focusing on a single fauna group (*Plecoptera*, *Ephemeroptera*, *Tricoptera*), fish, interstitial hyporheic fauna, plankton, as well as interstitial sediment granulometry and hydrochemical data gathered during the last thirty years shows a continuous degradation of water quality and aquatic and riparian habitats in many river sections (Braioni, 2001).

In principle, the wastewater treatment plants in each municipality along the Adige course can support the water quality of the river and its biota. However, many waste-water treatment plants along the river, which is regulated with respect to the flow for hydro-electric and irrigation uses and canalized with respect to the bed and banks and excavated, are not compatible with the needs of drinking water supply in the plains, where the natural self-purifying capacity is reduced and sediments are contaminated by the conveyance of polluted material from upstream and from tributaries (Braioni and Ruffo, 1986; Duzzin, 1986; Pavoni et al., 1987). Fragmented jurisdiction institutions and agencies in charge of controlling, planning, managing and servicing did reduce some environmental damages but were not able to safeguard the resources comprehensively.

9.5.3 Objectives of the case study

Classification and assessments of Water Bodies in the Adige River system

New concepts of environmental sustainability considering the broadest historical, cultural and socio-economic contexts, have re-evaluated the relevance of river and riparian areas in planning. They have become substantial elements of reference, essential for the recognizability, readability and importance of urban landscapes, fundamental for the ecological stabilization of industrialized countries and a backbone of global ecological networks.

This has been confirmed by Italian regulations, EU Directives and the Recommendations of the European Landscape Convention and the Italian Urban Codes, which unify all Italian legislation concerning the landscape (D.L., 1989; D.Lgs, 1999; 2004; Directive 2000/60/EC, Paour and Hitier, 1998). The criteria and environmental parameters thus taken into account by different disciplines have become decisive in the management and planning of the territory. Italian rules prescribe periodic monitoring to be made by regional and provincial environmental agencies. These agencies, at the same time as staff of Laboratories of the Agrarian Institute of S. Michele, the Natural Science Museum of Trento and the Biological Laboratory of Laives, have also been involved in the ecological monitoring at structural and functional levels in coordination with the researchers from the Padova, Bologna and Venice universities (Braioni, 2001).

Such analyses conducted by Geographic Imaging System (GIS) and digital Cartography of the National River Basin Authority of the Adige river (ATI CSR Presspali, 2000) have helped to define the following information for water bodies (Braioni et al., 2000; Braioni et al., 2002):

(1) classification of the surface water quality as requested by the Italian Norms (D. Lgs, 1999)to be done continuously along the river
(2) classification of the surface water quality according to use – drinking water source, fisheries, irrigation, bathing, continuously along the river (D. Lgs, 1999)
(3) evaluation of the ecological status along the entire river course as required by the Water Framework Directive (Directive 2000/60/EC)
(4) assessment of biological, functional and structural as well as ecological, environmental and landscape qualities of the Adige River fluvial corridor (Braioni et al., 2000; 2002a).

These results indicate that monitoring water bodies with respect to their classification for various uses and defining their ecological status is important for identifying water quality objectives to be achieved. Still, these monitoring methods do not offer enough data to show the possible structural and management interventions necessary to reach the given quality objectives. The procedures proposed in the integrated management and planning of the ARFC system (which also uses multiple ecological indices and functional evaluations) provide a synthesis and a representative view of the structural and functional aspects of the river ecosystem, which planners can easily interpret.

Environmental Planning approach

The approach used in this case study is based on environmental planning that can be defined as an analytical design process, by which the public subject intends to govern a specific geographical space selecting all the possible variables which constitute the eco-sphere and represent best the complexity of the environmental organism. As a matter of fact, the purpose of the environmental planning approach consists in improving the global quality of the environment through developing a better balance between anthropogenic impacts and natural resources. Thus, it is a step forward and an improvement in sector planning, considering economic, social, and ecohydrological (prevalence of the biological, chemical, ecological and physical variables) issues, landscape variables (prevalence of emotional and perceptive variables), and urban planning (prevalence of territorial organisation variables).

In the *Best Practices in River Basin Planning, Guidance on the Planning Process* (CIS WFD, 2003), ten different types of integration issues are identified. The proposed method is used to integrate, in the planning process, the ecological and biological quality of the river, ecological-naturalistic-environmental landscape quality of the riparian areas, and fluvial corridor land-uses. In this model, analyses produced by different disciplines are inter-related both in the data collection phase and in the final evaluation, thus determining trans-disciplinary cognitive processes and a continuous comparison among different models and analytical methods. This is also an integral part of the Recommendation 40 in the Draft European Landscape Convention (Paour and Hitier, 1998).

9.5.4 Project results

There are many studies assessing the results of the approach applied here (Braioni et al., 2000; 2001a; 2001b; 2002; 2005; 2006). Two examples of the thematic maps and the multi-disciplinary maps (The land-use map and the quality, degradation and risks map) are presented in Figures 9.15 and 9.16. The latter evaluates all the related disciplines' analyses by identifying and sharing the elements of quality, degradation and risks among urban and landscape planning components, vegetation components, biological and ecological components, physical and chemical components, and morphological components.

This integrated assessment method of environmental complexity is essential to identify the actions and interventions of ecosensitive and sustainable planning for the future. The Possible Land-uses Map presents the future usage of the river–fluvial corridor system. General planning directions and the related abacus were defined for each of the eleven sub-sections.

According to the integrated interpretation of the multi-disciplinary results acquired for the eleven sub-sections of the ARFC system, forty-six types of interventions have been identified for the abacus attached to the cartography of the tables in the Potential Uses Map.

9.5.5 Project implementation

The model of integrated management and planning of the Adige River fluvial corridor system (Braioni et al., 2002a) shown in Figure 9.17 is aimed at achieving the following objectives:

- Supporting the urban planning discipline with the tools of sustainable environmental management of the river fluvial corridor system

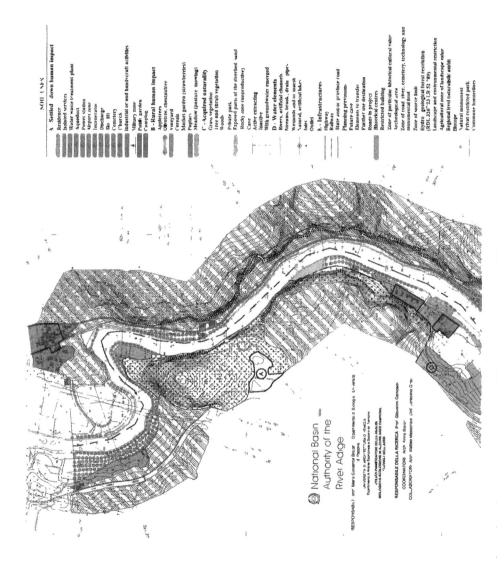

Figure 9.15 Land-uses in Sub-section 9 (See also colour plate 17)

Source: Braioni, 2001.

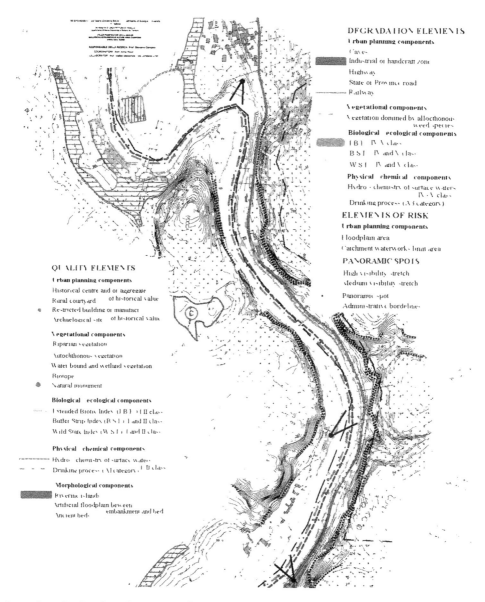

Figure 9 16 Quality, degradation and risk assessment of Sub-section 9 (See also colour plate 18)
Source: Braioni, 2001.

- Using the analyses of biology, ecology, hydrology, and landscape disciplines for planning
- Proposing planning actions for re-qualifying, re-naturalizing human impacts and rehabilitating the river-fluvial corridor system.

The model is flexible enough to be interfaced with other disciplines. The steps of the first phase are as follows:

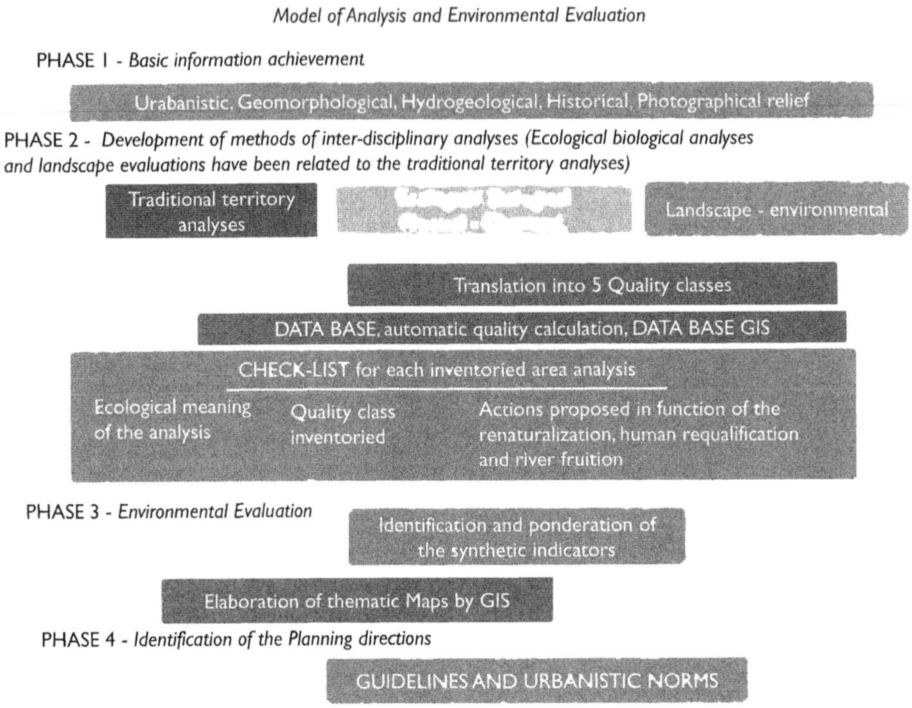

Model of Analysis and Environmental Evaluation

PHASE 1 - Basic information achievement

Urabanistic, Geomorphological, Hydrogeological, Historical, Photographical relief

PHASE 2 - Development of methods of inter-disciplinary analyses (Ecological biological analyses and landscape evaluations have been related to the traditional territory analyses)

Traditional territory analyses

Landscape - environmental

Translation into 5 Quality classes

DATA BASE, automatic quality calculation, DATA BASE GIS

CHECK-LIST for each inventoried area analysis

| Ecological meaning of the analysis | Quality class inventoried | Actions proposed in function of the renaturalization, human requalification and river fruition |

PHASE 3 - Environmental Evaluation

Identification and ponderation of the synthetic indicators

Elaboration of thematic Maps by GIS

PHASE 4 - Identification of the Planning directions

GUIDELINES AND URBANISTIC NORMS

Figure 9.11 Implementation Model (See also colour plate 19)

Source: from Braioni et al., 2001a.

(a) Selection of Sub-sections significant for all the related disciplines regarding the availability of historical data

(b) Standardization of the sampling protocols, and the timetable of the sampling compatible for collection of the data along the river continuum

(c) The choice of the most appropriate indicators and indices to evaluate the quality of compartments and components, aggregated, disaggregated and provisional capacity (Braioni et al., 2001a; 2001b; 2002a; 2005a; 2005b; 2006)

(d) Collection of the territorial data usually analysed in urban planning

(e) Surveys of the environmental aspects such as landscape, ecological, historic and urban related ones in order to assess the evolution of urban processes in relation to biological-ecological backgrounds

(f) Integrating urban analyses with all other disciplines. The relationship between humans and nature will shift from impact to integration. This approach will require an interdisciplinary contribution.

The interaction barriers among related disciplines are assumed as they are overcome in the second phase:

(a) Methods of transferring quantitative results to the trophic-functional biological-ecological, botanical processes as in qualitative classification norms with 5 classes (Braioni et al., 2002; 2005a; 2005b)

(b) Application of checklists for the synthesis and compilation of ideas and research coming from each discipline expert (Braioni et al., 2005a; 2005b)

(c) Transferring the results obtained in the above stages to the GIS and actual legislation, the WFD, addressing the evaluation of the biological-ecological quality, territorial data and the environmental landscape ones

(d) Coding modalities of data on digital maps maintained by the Administrative Authority.

In the third phase, specific disciplinary themes are elaborated, and in the fourth phase, the abacus planning is defined in multi-disciplinary and inter-disciplinary themes (Braioni et al., 2005a; 2005b; 2006), which can include norms, planning or management directions, and types of interventions. At this stage, data are graphically represented by maps, plans, sections and sketches of the present state and future proposed land-uses of the area.

The 'procedural/action abacus' is a double-entry grid containing the functional classification of the main types of actions following the system analysis correlated to their possible location in the space and time frame. This method allows for controlling possible overlaps of the contradictory norms and the continuous revision of the general planning with successive programmes of actions. The abacus model represents a planning approach that is particularly useful in regions with highly complex comprehensive variables (environmental quality, wood patrimony, landscape structure, areas characterized by different human impacts, etc.), and, at the same time, it overlaps other planning actions in use. The approach should proceed with a wider perspective of strategies that take into account the continuous communications among territory context/ecosystem functions/laws in order to perform human interventions in a more sustainable and manageable way. The procedural Abacus can support both a vision/overall planning and the accomplishment of immediate actions. So it is possible to operate planning and process. By its own nature of systemic analysis and sets of laws and methodological indicators, the abacus is to be considered as an instrument for facilitating progress by featuring a constant open elaboration (Braioni et al., 2006).

9.5.6 Stakeholders and their roles in the project

The research addressing integrated management and planning of the properties of the Adige River Fluvial Corridor System was financed by the National Basin Authority of the Adige River in support of the Adige Basin Plan. Local studies are carried out with funding from the Veneto Region, BZ and TN Provinces. They are coordinated, integrated and complementary to those managed by the NBA of the Adige River. The Biology Department (University of Padova) collaborated with the LASA Department (University of Padova), the Evolutionary Biology Department (Bologna University), DAEST-IUAV, the Biological Laboratory (Laives, BZ), IASMA (TN), the Natural Sciences Museum (TN), and the Environmental Protection Agencies of the BZ, TN, VR, PD, RO, VE Provinces.

9.5.7 Conclusion and recommendations

Coordination and inter-calibration are necessary among all institutions carrying out the monitoring of aquatic habitats. It is possible to monitor the quality of surface water, groundwater and rivers according to the designated Italian standards and then assess in detail the critical sections of aquatic habitats. Periodic monitoring of the Adige river fluvial corridor system by functional and complex indices is indispensable to follow the evolution of the components and compartments of the system. Quantitative and functional biological analyses (arboreal and erbaceous vegetation, phytosociological analysis, floristic analysis of the banks and riparian areas, riparian fauna, biotic functional processes in the river-bed including colonization and litter breakdown in organic and inorganic substrata, translated into fives classes), can be transferred onto maps, preserving a high level of precision and sensitivity in quantifying both the quality of the environmental conditions and the integrity of the ecosystem processes. Complex Indices, such as Wildlife State Index, Buffer Strip Index, and Environmental Landscape Index, which were applied along the riparian areas and fluvial corridors (Braioni et al., 2001a; 2001b; 2005a; 2005b), were very useful for the integration of biological, ecological, and environmental landscape indicators with the territorial urban planning data. Their qualities, aggregated or disaggregated, can be used in multi-disciplinary maps. Therefore, in every single sub-section, the pressures of the driver, the state, the impact and the responses can be underlined on the basis of objective data (Braioni et al., 2006).

In the areas of historical and environmental significance, and in those that biology and botany experts indicated as having natural value, particularly if protected, such as Sites Natura 2000 (Sub-sections 1–3, 6–11) or areas of Natural Parks (as for example the urban fluvial Park in the Verona City), and the degraded areas, the construction of a process of attention to the territory transformation by progressive checks on different levels of actions and sharing is indispensable. In fact, the procedural/action abacus helps to control the possible overlap of the contradictory norms and the continuous revision of the general planning with successive programmes of actions as described above.

9.6 ASSESSING STREAM BIO-ASSIMILATION CAPACITY TO COPE WITH COMBINED SEWER OVERFLOWS, LYON, FRANCE

Pascal BREIL and Michel LAFONT

Cemagref, 3 bis quai Chauveau, 69336 Lyon cedex 09, France

Stream bio-assimilation capacity is assessed in a developing peri-urban area on the west side of Lyon, France. The total population of Lyon is 1.6 million (2000) and the climate is mild continental with Mediterranean influences. Perennial and intermittent small streams with porous sediment beds and high stormwater and groundwater exchange capacity represent ecosystems very sensitive to pollution. Threats include poor water quality, physical habitat degradation and flood hazard. This study's objective is to adapt Combined Sewer Overflows (CSOs) to the bio-assimilation capacity of the receiving streams (see Figure 9.8). Major actions consist of identifying factors that control bio-assimilation capacity, and exploitating the enhancement of these factors for improving the state of water resources in the catchment.

9.6.1 Background

The City of Lyon is facing rapid urban development. Its west-side catchment is a hilly land drained by a dense network of streams and seasonal tributaries. As presented in Figure 9.19, the total water demand is greater than the natural water input to the catchment, and natural water resources are unable to fully support the needs of the present population. A complementary supply of potable water comes from the alluvial

Figure 9.18 **Yzeron River, Combined Sewer Overflows (See also colour plate 20)**
Photo credit: P. Breil.

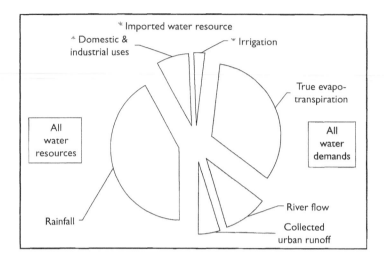

Figure 9.19 Western suburb of Lyon city: Unbalanced water resources and demands

plain of the Rhône and Saône rivers, on whose confluence the old Lyon city centre was built. A similar situation is often observed in peri-urban areas of large cities of the word.

During the last twenty-five years, the urban area of the catchment studied (Yzeron) has grown from 20 to 25% of the total area (150 km^2) and is expected to reach 50% in 2030, with serious impacts on water resources. For this ultimate development, hydrological simulations reveal that the peak flows of small floods will be about ten times higher than in the present situation. Such a result is confirmed by the synthesis of data from multiple sites (Hollis, 1975). Most small floods result from CSOs, which means that urban water pollution reaching the stream system should greatly increase at the same time. Such impacts should also affect the benthic sediment ecosystem (Grapentine et al., 2004).

On the other hand, the annual volume of water flowing in the main sewer pipe at the outlet of the watershed is presently composed of 44% domestic wastewaters, 14% stormwater and 42% groundwater. Given the amount of water flowing in the Yzeron river at the outlet (see Figure 9.19), the combined sewer network conveys about a quarter of the natural rainfall and groundwater flow from this watershed. Urban development is then expected to deliver more pollution to the stream during rainfall events and to withdraw good quality water that feeds the river during dry weather periods.

Seepage of good quality water into the sewer pipes results from the degradation and neglect of maintenance of this thirty-year-old sewer network. Rehabilitation of such an infrastructure is costly, and the replacement option seems to reach its sustainable limits when addressing future peri-urban development. At the same time the limited flow capacity of the old sewer network also increases the number of CSOs. Thus, best management practices need to be proposed and integrated in the urban planning (see Chapter 4).

9.6.2 Key aquatic habitat issues in urban water management

Aquatic ecosystems consist of some perennial but mostly intermittent streams. Running waters are oligotrophic as a result of a granite bedrock covered by a thin layer of acid soil. Stream bottom sediment is dominated by sand, with thickness reaching more than 1 m in depth in some places, but it greatly varies along the water course. Prevailing hyporheic and benthic interstitial fauna are present in the bed sediment. They are composed of species tolerating dry periods and disconnections between surface water and the otherwise connected groundwater. Porous sediments offer a high capacity for exchanging surface and ground water, resulting in a very active ecosystem which is able to bio-assimilate nutrients (Breil et al., 2007), while at the same being time highly sensitive to pollution. Presently, there are well preserved stream corridors extending from the upper rural part to the lower urbanized part of the watershed. Such a layout avoids direct pollution inputs from agricultural (cultivated) areas and from direct urban runoff (see Chapter 7). However, at the same time, about forty combined sewer overflows are located along the main watercourses, including some on tributaries. The preservation of sensitive stream corridors and high resilience of the aquatic ecosystems supported by these habitats, with respect to coping with pollution, is a key issue in managing CSO impacts in this catchment.

9.6.3 Objectives of the case study

Streams and rivers use self-purification processes to improve water quality, if the habitats and biota are properly managed, without impairing their natural bio-assimilation processes (Zalewski, 2000). This is a natural potential that can be quite reactive and intense in sandy bottom streams like in the Yzeron catchment. Although so far these stream properties are not yet used extensively, research demonstrates their existence and usefulness (Jones and Mullholland, 2000; Hancock, 2002; Hancock et al., 2005; Lafont et al., 2006a; 2006b).

Considering the water management issues in the study area, and the potential benefits resulting from the understanding and proper management of the aquatic habitats quality, the objectives of this case study were as follows:

- risk identification based on comparing CSOs loads and bio-assimilation capacity of the receiving streams
- defining better locations for CSO outfalls with respect to lessening the stress on aquatic habitats
- developing local intervention measures for artificial enhancement of bio-assimilation capacity in stream corridors (Kasahara and Hill, 2006)
- assessment of the stream bio-assimilation capacity contributing to the overall increase of the reliability and resilience of the urban water cycle, with respect to increased water demands and associated threats, and increased variability of water resources as expected from global climate change.

All of the above measures are steps towards achieving the major goal of the case study, which is increasing the urban water cycle reliability in providing water supply and ecosystem services to the local community.

Before implementing the plan of the bio-assimilation potential enhancement, improving sewage quality prior to discharge into streams and rivers is a priority. It must be done by reducing pollution inputs to receiving waters by means of Best Management Practices (BMPs) linked to CSO controls (see Chapter 4). However, these objectives cannot be fulfilled in a single step, and consequently, only the two first tasks have been addressed in this case study.

9.6.4 First steps of implementation: assessment of bio-assimilation capacity

The first step in implementing this project was to identify physical determinants of the bio-assimilation capacity along the stream network of the Yzeron catchment. This was performed over a three-year research programme supported by the Yzeron river contract (which combines thirty districts), the Lyon City Council and regional-level funding. Some research projects funded by the EU and national sources also contributed to producing these results.

Characteristics of the physical determinants were defined by using the premise that in porous media the bio-assimilation capacity depends on the active exchanges between surface and groundwaters (Jones and Mullholland, 2000; Hancock, 2002; Hancock and Boulton, 2005; Hancock et al., 2005; Breil et al., 2007). In the case of the Yzeron streams, the hyporheic corridor is several metres wide and not more than a metre deep. This corridor is considered to be the major site for degradation processes in the river.

During the research project, several field sites were instrumented with flow monitoring stations, water samplers for water quality analysis, and piezometers for groundwater table gradient assessment. Tracer studies were used to identify flow paths and origins, and were complemented by regular biological sampling in the benthic and hyporheic substrates of the streams. These data allowed for observing active exchanges of water between the water column and the bed sediment and demonstrated their dependence on local geomorphic forms like pools, runs and rifle sequences and meanders (Breil et al., 2005; Breil et al., 2007). A potential hydric exchange rate was assessed using a filtering potential of stream substrates and banks (Lafont et al., 2006c).

A geomorphic typology of the valley and stream corridor features was performed using a multivariate analysis of information layers on geology, soil structure and texture and topography by the way of a GIS software (Schmitt, 2007). The resulting hydro-geomorphic typology was validated by field observations at fifty sites. Channel characteristics were statistically assessed along several sections of each hydro-geomorphic type. Nine classes were finally retained to describe the overall stream network. Extreme classes corresponded to steep gradients on bedrocks with boulders and blocks, and to flat gradients with sand, gravel and pebbles. Frequencies and lengths of pool, run and rifle units were also integrated in the typology. Whenever an overall upstream-downstream trend exists in the class type succession along the main streams of the watershed, a great variety of situations is found in sub-catchments.

Dominant types of land-water connections to the stream corridor were assessed using a distributed hydrological simulation model (Gnouma, 2006). Contrasted situations were defined considering soil properties, gradients of land adjacent to the stream corridor, land-use, and perennial and seasonal stream reaches. It resulted in three types of land-stream channel connectivity: dominant infiltration, dominant runoff and intermediate.

9.6.5 Project results

Considering the variety of types of surface and groundwater connections and the hydro-geomorphic classes along the stream network, it was proposed to assess the

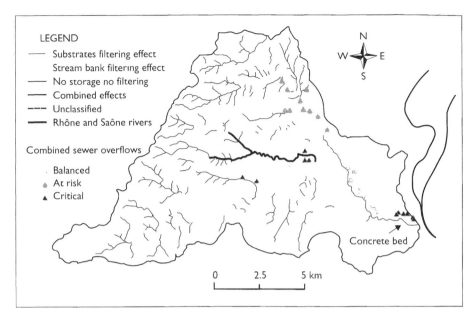

Figure 9.20 Results of the of bio-assimilation capacity assessment in the Yzeron Catchment: Stream network delineation of the filtering effect as a determinant of the bio-assimilation capacity; CSO structure (outfall) locations, and classification of habitats as balanced, at risk, and in critical situations

filtering potential of the stream corridor by a substrate filtering effect, stream bank filtering effect, no storage or filtering effect, and combined substrate-bank filtering effect. It was assumed that the bio-assimilation capacity was correlated with the filtering potential, classified in a descending order as: combined or full stream corridor based potential, substrate based, bank based and no filtering effect. The latter included such stream features as a concrete channel or a bedrock channel. The results of this analysis are summarized in Figure 9.20.

Superimposing the locations of CSO structures on the map of estimated potential for filtering made it possible to rank the potential impact of CSOs on aquatic habitats, especially those at risk and in critical situations, as presented in Figure 9.20.

In the next step, the potential for bio-assimilation capacity is to be assessed in terms of nutrient fluxes. So far, it was based on biotic functional traits, revealing the importance and direction of flow exchanges (Lafont et al., 1996; Lafont, 2001) and on conventional biotic indices, such as a general biological quality (invertebrate index), biological sediment quality (oligochaete index) and biological water quality (diatom index). Field observations reveal that at the sites characterized by good filtering capacity, the combined sewer overflows impact is smaller. Periodic biotic sampling also revealed that reaches with a good potential bio-assimilation capacity, and thus also a good resilience capacity (see Chapter 2), can recover quite rapidly, even when under urban impacts (Weyand and Schitthelm, 2006). Thus, adaptation of the frequency and volume of CSOs to these natural properties of the catchment seems to have a realistic potential to balance human pressure on streams and to contribute to achieving a good ecological status, as requested by the WFD.

The results also show that CSOs structures should be moved or spread along aquatic habitats considering the differences in their bio-assimilation capacity. It was observed that in some sections of the Yzeron stream network, characterized by efficient bio-assimilation capacity, CSO discharge loads, corresponding to a load from 1,000 people, can be adequately bio-assimilated over a distance as short as 100 m of a stream. This happens without any noticeable ecological degradation, but with cyclical biomass increases in living species in the bed sediment of the stream. On the contrary, sites receiving high volumes of CSOs exhibit important accumulation of organic matter in the stream bed sediment, with prevailing anaerobic conditions. These are polluted conditions that can represent a health risk when mixed with surface waters, such as during floods. The best CSOs management should avoid such situations.

9.6.6 Stakeholders and their roles in the project

The Yzeron River contract contains a series of actions to be implemented by a number of stakeholders directly involved in this process. These actions were agreed upon and planned during a four-year study as follows: list main sources of impairment, define corrective actions, and estimate financial budget. State, regional and local administrations contribute to the project funding. The first objective of the river contract is to rehabilitate and protect the water quality in streams. The research central role is to set up a series of performance indicators and design a field observatory to monitor the effect of the implemented measures. The objective is to get feedback on the efficiency of these actions regarding their initial objectives: water quality improvement, flood control and drought control.

9.6.7 Conclusions and recommendations

The western section of Lyon has, to a certain extent, resisted intensive urban development for several decades, mainly due to the lack of flat lands for development. It is by now experiencing a rapid urban growth, which reveals the inadequacy of existing sewage infrastructures. But the costs of upgrading and rehabilitating the sewer network are high, and the effectiveness of such measures in the coming decades has not been demonstrated. Moreover, rehabilitation costs should be included in the renovation of a sewer network. The considerations of sustainable development are forcing decision-makers and managers to find new alternative solutions to be combined with the existing infrastructures. Solutions based on natural processes taking place in aquatic habitats are under consideration, as they provide low-cost treatment facilities and sustainable solutions, if managed according to their natural needs for water and flow variability (Fatta et al., 2002). The education and training of water managers is essential for achieving wide acceptance, understanding and success in the implementation of new ways of thinking, and in the longer perspective, should be cost-beneficial for all citizens.

A better understanding of aquatic ecological processes should provide a basis for improving the ecological resilience of aquatic habitats with various features, and protect them from impacts resulting from land-use changes and climate variability.

9.7 OPTIMIZATION OF THE RIVER HYDROLOGICAL REGIME TO MAINTAIN FLOODPLAIN WETLAND BIODIVERSITY, LOBAU BIOSPHERE RESERVE, VIENNA, AUSTRIA

Georg Albert JANAUER

Department of Freshwater Ecology, Section of Hydrobotany, University of Vienna, Althanstrasse 14. A - 1090 Vienna, Austria

The Lobau UNESCO Biosphere Reserve case study encompasses the Danube River floodplain located in a sub-continental climate, within the limits of the City of Vienna, Austria, with 1.6 million inhabitants (as of 2000). Aquatic habitats addressed by this case study include floodplains, wetlands and riparian corridors, rich in flow sensitive biota. Threats include the metamorphosis of water bodies through wetlands to dry land due to siltation, as well as flood impacts caused by oxbow revitalization. The major goal is to sustain and enhance biodiversity and aquatic plants ecosystem services, and major actions concern ecohydrological planning, legal protection and administrative balance of stakeholders.

9.7.1 Background

Environmental context

The Lobau is part of the historical floodplain area of the Danube River (see Figure 9.21 and Figure 9.22). Today floodplain lakes, wet ditches and riparian wetlands still exist.

Figure 9.21 Kühwörther Wasser oxbow, Lobau wetlands, Vienna (See also colour plate 21)

Photo credit: G. Janauer.

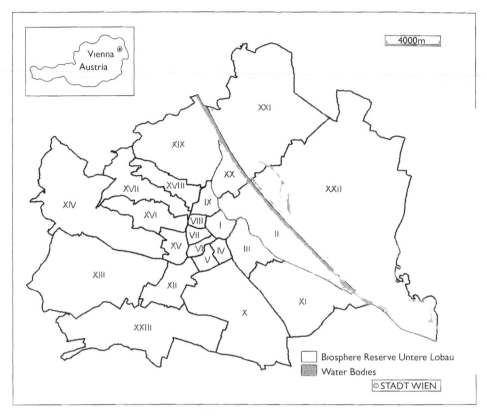

Figure 9.22 Location of the Lobau Biosphere reserve in Vienna (See also colour plate 22)

Source: http://www.wien.gv.at

Table 9.2 Basic climatic characteristics of Vienna

parameter	value	Period
Precipitation [mm]	668	average 1990–1999
	998	max – 1941
	404	min – 1932
Evaporation [mm]	550	average – 1998
Temperature [C]	38.3	max, 08.07.1957
	25.8	min, 11.02.1929

Source: Vienna, Hohe Warte.

Their ecological value is enhanced by the fact that they form a last refuge for aquatic organisms having lost their former habitats in the adjacent landscape, which is dominated by urban development and agricultural areas.

The catchment area of the Danube in Vienna is approximately 100,000 km², with river discharge (mean 1,900 m³·s⁻¹) ranging from min 650 m³·s⁻¹ to max 15,000 m³·s⁻¹ (calculated for the 1,000-yr return period). Major climatic characteristics are given in Table 9.2. The water quality is at, or close to, the natural conditions expected for a river of this size; e.g., phosphorus concentrations and the composition of the inverte-brate fauna in the river sediment classify its condition as oligo- to mesosaprobic, which corresponds to the carbon load under natural conditions. Of course, this is the result of rigorous sewage treatment throughout the catchment, which includes tertiary treatment.

The Lobau floodplain ensemble includes still waters and riparian forests remaining after river regulation in the late nineteenth century. The Lobau is separated from the river by a levee, but passage of seepage water is uninterrupted and floods flow in an out through a small opening in the levee at the downstream end of the area. Prevailing still water conditions and 130 years of siltation trigged natural succession processes, which resulted in enhanced development of reeds and other wetland vegetation and considerably reduced cross-section and shape of the water bodies. Abundant macrophytes and the strong groundwater influx from the river keep the water quality in the floodplain lakes at a high level.

Historical background

Prior to the regulation of the Danube in Vienna, the Lobau was one of many islands in the natural river corridor. Discussions on how to cope with the threat of Danube floods lasted for more than two centuries (Michlmayer, 1997). Auxiliary works started in the late 1860s, and straightening of the rivercourse and cutting of the main river arms and meanders, was finished by 1880. The regulation instantly started the deep-ening of the river bed. Several oxbows and former side-channels form habitats for a multitude of aquatic plants today, most of them claimed in the Red Lists as endangered species (Schlögel, 1997).

In 1954 a flood with about a 100-year return period damaged the flood protection dams and flooded the northern and southeastern parts of Vienna. This triggered the search for an 'ultimate' flood protection system, which consists of the 'New Danube' flood relief channel parallel to the main river bed, and the 'Danube Island' between these two water bodies. Intensive biological investigations started in 1985 under the lead of a group of Viennese limnologists (Janauer et al., 1985), who co-operated with architects and engineers to realize an ecohydrological approach. The Vienna Biotope Survey, which included all surface waters in Vienna (Lobau area: Janauer, 1982) was an essential knowledge base for the planning team.

General characteristics of the urban area

Vienna is located in one of the most prosperous regions of Europe. This situation forms the background of all socio-economic and environmental aspects related to the city and the aquatic habitats. Over the last decades, large public housing complexes and individual housing have expanded into the areas to the north and northeast of the Danube River.

The residents live on an area of about 420 km^2. The city is located between the Eastern limit of the European Alps, which is formed by a chain of small mountains, and a wide geological basin towards the east and south. Built-up areas occupy about 14,000 ha, the size of green areas (including agricultural land) is approximately 20,000 ha, water bod-ies occupy around 2,000 ha, and traffic and communication corridor areas amount to about 6,000 ha.

Despite regulation and the intensive development of the city, the Lower Lobau area was declared a UNESCO Biosphere Reserve in 1977, Natural Sanctuary under Viennese law in 1978 and incorporated into the Danube Floodplain National Park under the Austrian Federal Law in 1996. It is also a protected area according to the RAMSAR Convention, and under the NATURA 2000 Directive of the European Communities.

Urban water management

The main source of potable water of premium quality and sufficient quantity (average daily consumption is $390,000\,m^3$) originates from two high alpine sources located in the carbonate rock Eastern Alps. A reservoir fed by a little river and filtrate from the Lobau wetlands, as well as from a large groundwater reservoir in the Vienna Basin are auxiliary sources in the case of water shortages.

Within the urban agglomeration 98% of Vienna and an additional 62,000 PEs (person equivalents) in smaller towns, as well as 18 streams, are connected to the gravity sewer system. The remaining wastewater is collected in mostly sealed septic tanks and disposed of to the main sewers (2,158 km of sewers across the city). Stormwater is fully retained in sewers by automatic devices and then fed to the central sewage purification plant, with a maximum capacity of 4 million PEs. The plant treatment train includes phosphorus stripping and nitrogen elimination, with an overall efficiency more than 95%. Toxic substances are pre-treated at the source, and the effectiveness of the respective legal regulation is strictly controlled.

Responsible authorities

Urban water management, including drinking water supply, waste and rainwater treatment, water engineering, and wetland conservation are central issues for several municipal administrative units. Federal ministries are responsible for international navigation on the Danube, implementation of the EU WFD, and national water-related laws. Hydro-electric power production has been privatized. Stakeholder interests, however, still clash and some are focused on the future of the Lobau wetlands.

9.7.2 Key aquatic habitat issues in urban water management

Current key issues include finalizing flood protection for Vienna, reducing sedimentation in the Lobau water bodies during the recession of floods, re-establishing higher hydrological dynamics in the wetland in an attempt to restore at least some of the former character of the natural floodplain system, while saving existing biodiversity at the same time. Conflicts exist over the future role of drinking water wells in the Lobau. In the Danube reach downstream of Vienna, the river bed incision (1.5 m in the Vienna reach between 1875 and 1990) and the corresponding sinking of the groundwater level were caused by straightening the once meandering river course: these issues need ecosystem friendly solutions.

The 21-km-long artificial watercourse, called the New Danube, is located in parallel to the main river channel. It diverts floods, but its two impoundments are centres for leisure activities in combination with the 'Danube Island'. Even the maximum calculated discharge of the Danube in Vienna ($15,000\,m^3s^{-1}$) should safely pass the city today. In the New Danube abundant aquatic plants regulate the nutrient regime and compete with microscopic algae, thus preventing possible toxic algal blooms (Janauer and Wychera, 2000).

Figure 9.23 Schwarzes Loch floodplain lake in the Lobau Biosphere Reserve (See also colour plate 23)

Photo credit: Strausz, 2006.

9.7.3 Objectives of the case study

The 'UNESCO Biosphere Lobau' was created to save the ecological values of a wet-land and floodplain waters ensemble of international importance, indicated by its high biodiversity and a set of still water species of plants and animals, which have found a last refuge in this part of the Danube River Corridor, as seen in Figure 9.23 (Janauer, 2000). The special combination of moderate to intermediate flood pulses and a still effective groundwater connection with the main river channel dominate the hydro-logical regime in the Lobau area.

Final steps in flood control needed for the river reach in Vienna, and at the same time saving the high quality of potable water reserves in the groundwater system of the Lobau wetlands, must be merged with the intrinsic ecological values of the area.

Planning the ecological future of the Lobau and respecting the life in its water bod-ies is not a single-purpose activity, as in the past. Aiming at an ecological situation that reflects the conditions prior to the river regulation calls for water flowing through the Lobau. This certainly is a situation not in favour of many of the still water organisms surviving there today.

The prime task of the 'Demonstration Site Project Lobau' is therefore the intensive col-laboration among all stakeholders and searching for solutions that will do the following:

(i)   fulfil the need to sustain existing biodiversity as much as possible
(ii) try to find new refuges for the established still water species of flora and fauna
(iii) not to interfere too much with the need for flood protection and the wish to re-establish more dynamic flow conditions in the main channel system of the floodplain.

9.7.4 Expected results

Today relevant activities have been started by the Demonstration Site Project leader and it is likely that solutions for the above mentioned problems can be found by applying ecohydrology principles (Zalewski et al., 1997; Zalewski, 2000).

Keeping a reasonable level of biodiversity, especially of the aquatic and wetland vegetation, which helps to sustain, and even enhance the water quality by reducing algal bloom is possible by finding – and even creating – new refuges for still water biota. This in turn will support the intended hydrological restoration – and will be less single-sided in a general ecological context. Upon success in sustaining biodiversity and intensifying hydrological dynamics, existing environmental education will become even more attractive. In its course, cooperation with the Danube National Park management, the Austrian UNESCO Commission and the participation of NGOs should also be intensified.

The scientific assessment of biodiversity and ecological functioning of wetlands and floodplain water bodies will allow for modelling habitat conditions, as well as for preferences for protected and endangered plant species functioning as bio-indicators and umbrella species in the aquatic-terrestrial ecotone.

9.7.5 Project implementation and stakeholders

Flood diversion as a protective measure and the restoration of near-original flow dynamics and hydrological regime will exterminate flow susceptive aquatic plant and animal species. Therefore, refuge habitats must be found or created to sustain existing biodiversity of macrophytes in the best possible way. This, in turn, will control excessive algal growth in newly created water bodies. Phosphorus and nitrogen bio-accumulation and structural functions, which are key ecosystem services by preserved aquatic macrophyte vegetation, will be sustained even under new hydrological conditions (see Figure 9.24).

The following institutions are involved in the future development of the Lobau Biosphere Reserve: The Councillor for the Environment, Vienna, several municipal administration units (MA 45 – Water Engineering, MA 22 – Environmental Protection, MA 31 – Waterworks, MA 49 – Forestry), the Danube National Park Administration, the Department of Freshwater Ecology at the University of Vienna, the Austrian Academy of Sciences, the Austrian UNESCO Commission, and the Federal Province of Lower Austria. Stations monitoring water quality are operated by the Federal Office for Water Management, the MA 45 – Water Engineering, and the Austrian Hydro Power, Inc (AHP). Education and training is conducted by the Municipal Office, MA 22 – Environmental Protection, the Danube National Park Administration, the Austrian UNESCO commission, the Federal Agency for Water Management, and the Department of Freshwater Ecology.

This composition of the stakeholder group provides a fair chance to reach truly sustainable solutions for this ecological jewel, the Lobau Biosphere Reserve, which is a perfect example of a well managed wetland situated within an urban agglomeration.

9.7.6 Conclusion and recommendations

Far reaching changes are planned for the hydrological regime of the UNESCO Biosphere Reserve Demonstration Site Lobau in Vienna. Looking at the present status,

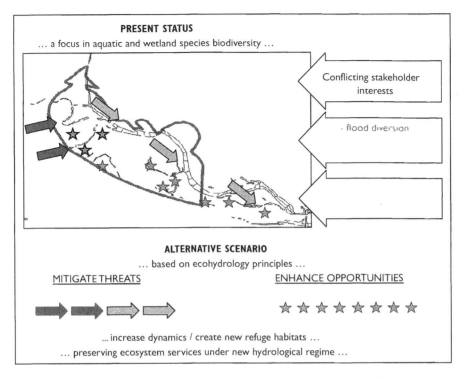

PRESENT STATUS
... a focus in aquatic and wetland species biodiversity ...

Conflicting stakeholder interests

· flood diversion

ALTERNATIVE SCENARIO
... based on ecohydrology principles ...

MITIGATE THREATS   ENHANCE OPPORTUNITIES

...increase dynamics / create new refuge habitats ...
... preserving ecosystem services under new hydrological regime ...

Figure 9.24 General concepts of Implementation Scheme (See also colour plate 24)
© Janauer. 2006.

the level of biodiversity and the quality of the floodplain ecosystem are still high, and therefore the area is still in compliance with the earlier aims, which led to the introduction of protection measures, since this site is a Biosphere Reserve, a Ramsar Site, a Natural Protection Area, and part of the Danube National Park. Of course, studies must be undertaken to detect in more detail the hot spots of ecological quality and to assess the vulnerability of the aquatic vegetation and to find refuge habitats for those species not adapted to a running water environment.

The first and most important recommendation is the application of integrated planning in developing the general strategy which must include the preservation of as much as possible of biodiversity and preserve ecosystem services in the wetland area (which also relates to the issue of the drinking water wells). Applying the principles of ecohydrological planning and having all stakeholders agree on the aims that can be reached form the basis on which to build the concept for technical measures needed to enhance floodplain dynamics and at the same time to save existing ecological values. If these principles are applied a balance between diverging stakeholder interests will be found – just as it was the case in all prior far-reaching decisions, which had to be made in the series of interventions in the hydrological regime of the Lobau area having started with the river regulation a long time ago.

9.8 INTEGRATED MANAGEMENT OF AQUATIC HABITATS: URBAN BIOSPHERE RESERVE (UBR) APPROACH FOR THE OMERLI WATERSHED, ISTANBUL, TURKEY

Azime TEZER

Urban and Regional Planning Department, Istanbul Technical University (ITU), Taskisla 34437, Taksim, Istanbul, TURKEY

The Omerli Watershed (OW) is the most important among the seven watersheds that provide drinking water to Istanbul, a megacity with over 10 million people (as of 2000). The OW is threatened by urban development, as is the case of other drinking water sources for the city. It faces the most acute, unplanned pressures of urbanization with serious impacts on water quality and biodiversity. The proposed Urban Biosphere Reserve (UBR) attempts to reconcile urban development, water quality and biodiversity conservation in a more sustainable way through integrated urban aquatic habitat management.

9.8.1 Background

The Istanbul Province is the most populated part of Turkey, experiencing a population growth of about 5% annually since the 1950s. The last two population censuses indicate

Figure 9.25 Aerial photo of the the Omerli Watershed (See also colour plate 25)
Photo credit: A. Tezer.

that the growth rate has been slowing down to around 3% annually (SIS, 2000). The province increased almost 700 percent in the built-up area as its metropolitan population grew from about 1,200,000 in 1955 to about 9,119,000 in 2000 (Tezer, 1997; Tezer and Kemer, 2004). Such drastic rates of growth of urban population and land area have inflicted widespread and devastating impacts on ecological life support systems of the region, especially on the structure of aquatic habitats and biodiversity. Over the last two decades, urban development in Istanbul has been taking place in and around its drinking water sources. Forests and water basins located to the north of the city have been experiencing considerable degradation since the 1980s, and the continuous 'building amnesties' encouraged illicit and unplanned developments extensively (see Figure 9.25 and Figure 9.26).

Extensive areas of grasslands and prairies on the European side and the heathlands on the Asian side were replaced by urban developments, with little attention to the habitat's characteristics and ecosystem services. Aquatic habitats especially have been seriously contaminated as a result of intense urbanization pressures, as in the case of the Kucukcekmece Watershed, which can no longer be used for drinking water supply. The other two freshwater catchments, Alibeykoy and Elmali, do not supply drinking water efficiently for similar reasons (Figures 9.26. and 9.27).

The Turkish Society for the Protection of Nature (DHKD) prepared a research report entitled 'Important Plant Areas (IPA) and Important Bird Areas (IBA) of Turkey' in 2005. The definition of IPA was proposed by the Planta Europa's Steering Committee as 'a natural or semi-natural site representing exceptional botanical richness supporting an outstanding assemblage of rare, endangered and endemic species' (Ozhatay et al., 2005). According to the report, 122 sites were designated as IPAs all around Turkey, with Istanbul possessing 10 different IPAs (see Figure 9.26) (Anon, 1999). Although this report does not mean that there will be a complete formal protection status declared for these areas, it points out comprehensively the importance of the natural character of the 122 sites in Turkey.

9.8.2 Key aquatic habitat issues in urban water management

Although Istanbul's cultural heritage is widely known, the uniqueness of its natural heritage has not always been as significant. The exceptional geographic location forms natural habitats and ecosystems with astonishingly diverse and rich floral character. Istanbul, with an area of $5,512\,km^2$, accommodates more than 2,000 native-vascular floristic and fern species in 5 different habitats: wetlands, grasslands, heathlands, coastal dunes and forests. This number is higher than that in the UK and the Netherlands' native-vascular species. There are eighteen species in the Bern Convention List to be protected and seven of them grow in aquatic habitats.

Except for the IPA/IBA No.4: Hills on the North Bosporus and the IPA/IBA No.7: Hadimkoy, Kemerburgaz Grasslands and Heathlands, the rest of the 10 IPA/IBAs in Istanbul are mainly aquatic habitats (see Figure 9.26 and Table 9.3).

The OW is the most important among the seven watersheds that provide drinking water to Istanbul. These seven watersheds cover $2,353\,km^2$ (42% of the total area of $5,512\,km^2$) of land area in the Istanbul Province. The OW, on the Asian side of Istanbul, provides the largest supply of drinking water (35%) and faces the most acute urbanisation pressures. It has the highest population (371,400 inhabitants in 2000) and

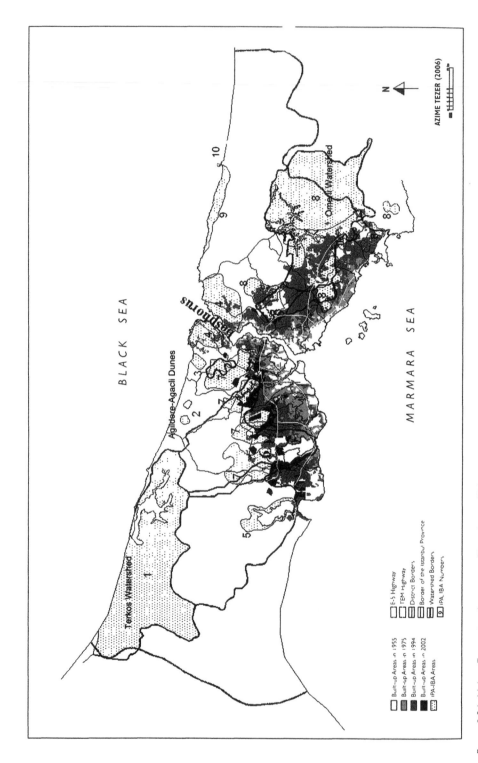

Figure 9.26 Urban Expansion, Important Plant Areas (IPA) and Important Bird Areas (IBA) in the Province of Istanbul (See also colour plate 26)

Source: Tezer, 2006.

Table 9.3 Aquatic IPA and IBAs as potential UBRs in Istanbul

Name and number of IPA/IBA Landscape (*)/Protection (*-*) Priority	Habitat/Biodiversity Characteristics	Threats
Terkos and Kasatura Forests (# 1) *A/Very Urgent*	– "Bern Convention" and "Globally Threatened Endemic Species List" – Best protected coastal dunes on the Black Sea coasts – Richest flora of freshwater lake ecosystems in Turkey	– Forestation – Agriculture – Water-pipeline network impacts on forests
Agil Dere and Agacli Coastal Dunes (# 2) *C/Very Urgent*	– Rich coastal dune biodiversity. – "Bern Convention" and "Globally Threatened Endemic Species List"	– Mining – Forestation with alien species to coastal dunes
Gumusdere Kilyos Coastal Dunes (# 3) *C/Critical*	– The second most important priority area of Turkey's coastal dunes of Black Sea coasts. – Important historical background of hydrobotanical species	– Residential development – Mining – Campus development – Recreational over-use
Buyukcekmece Lake (# 5) *B/Urgent*	– Compatibility to wetlands criteria of RAMSAR Convention – Species in Global Conservation Concern criteria	– Industrial developments – Residential developments – Uncontrolled hunting – Water pollution
Kucukcekmece Lake (# 6) *C/Urgent*	– Compatibility to wetlands criteria of RAMSAR Convention – Species in Global Conservation Concern criteria	– Dense built-up areas – Water pollution
Asian Side Heathlands and Omerli Watershed (# 8) *A (northern part)/Very Urgent*	– IBA – Last and largest heathlands of Eastern Mediterranean Region, quercus coppice and pinus negra forests – "Bern Convention" and "Globally Threatened Endemic Species List"	– Unplanned developments – Forestation – Agricultural expansion through heathlands
Sahilkoy-Sile Dunes and Forests (# 9) *A/B/Critical*	– "Bern Convention" and "Globally Threatened Endemic Species List" – Coastal dunes and *quercus* coppice	– Summer houses – Mining – Recreational over-use
Sile Islands (# 10) *A/Urgent*	– IBA – Coastal dunes – *Quercus* coppice	– Visitors threat on bird nesting areas

relates to the location of IPA/IBA shown in Fig. 9.26.
* Landscape Priority, C to A (A is the highest).
** Protection Priority of the area, as "critical", "urgent" and "very urgent"(very urgent is the highest priority).

(Excerpted from Tezer 2006)

Figure 9.27 Omerli Dam and neighbouring heathlands (See also colour plate 27)
Source: DHKD Archives.

built-up area, which have increased through unplanned and irregular developments since the 1980s. Recently, the area has become increasingly attractive for housing developments, because it is in a zone of low earthquake risk, and the availability of tempting housing loans and the proposed mortgage law. The major problems are the absence of integrated and macro-level planning strategies at the basin scale giving a proper consideration to aquatic habitat protection and management, the lack of coordination among major institutions, which direct planning implementation, the lack of public awareness of water and biodiversity protection issues, and finally the frequent changes in local plans and planning regulations according to the political tendencies of governments.

Since major consequences of biodiversity loss and environmental degradation have resulted from urban-based activities, it is necessary to examine the applicability of the Man and the Biosphere (MAB) Programme to define an urban-based biosphere reserve (BR) approach. Designing cities compatible with their ecological background will have a considerable impact on the establishment of sustainable landscapes at local level.

Conservation, development and logistic functions of BRs can be supported by urban green areas with a biological conservation hierarchy, starting with undisturbed ecosystems of a high biodiversity value through urban green systems and biodiversity improvement practices. The conservation function of BRs can be broadened to include degraded ecosystems and habitats to enhance biodiversity. In other words, the major goal of BRs, 'to protect unique habitats and ecosystems,' has to be expanded to the areas under pressure because of degradation caused by urban development.

An Urban Biosphere Reserve (UBR) may use the existing designations on biodiversity protection, such as protected areas along with urban development plans, and they have a potential of integration to establish ecosystem approach in local urban planning efforts. The Omerli Watershed has no protection status for its exceptional natural habitats at present, except limited 'watershed regulation' of the Istanbul Water and Sewage Board (ISKI). Although, ISKI has the authority to monitor and control developments inside the watershed borders, the Municipality of Greater Istanbul (MGI) controls the planning and implementation of urban development in the Province.

ISKI applies the watershed protection zones to control development and water quality. These are absolute-, short-, medium- and long-distance protection zones. Unfortunately, these are set by definite distances and not according to natural habitat characteristics. The proposed UBR for these areas will be effective in defining the 'natural habitat management' and also for the protection of biodiversity and water quality.

9.8.3 Objectives of the case study

The Statutory Framework of the World Network of Biosphere Reserves defines the departing point of UNESCO's MAB Programme in the Introduction section thusly: 'BRs are established to promote and demonstrate a balanced relationship between humans and the biosphere without distinctive definition of size, context and compounds of balanced interactions.' BRs promote the conservation of landscapes, ecosystems, species and genetic variation for fostering ecologically, socially and culturally sustainable development and providing logistic support for environmental education, training, demonstration projects, monitoring and research-related local, regional, national and global issues of biodiversity conservation and sustainable development. Urban Biosphere Reserves (UBRs) indicate compatibility with the context of the Statutory Framework and the Seville Strategy and may have crucial potential to promote urban sustainability in environmental, economic and social dimensions (MAB Urban Group Policy Paper, 2003; Dogsé, 2004).

The Asian Side Heathlands and the Omerli Watershed (IPA No.8) can be a proper node to start up the process of the management of UBR concept in Istanbul. The reasons for selecting this site are the 'very urgent' protection need of aquatic habitats and the water quality of the most important catchment with respect to Istanbul's freshwater sources; the existence of a significant amount of urban development pressures, and finally potential deterioration of globally and locally important aquatic and terrestrial ecosystems. The UBR concept will necessitate an integrated planning approach for research, development, implementation and monitoring phases. Additionally, the UBR concept has the potential for a multi-participatory and multi-objective character, with its integrated management structure. Therefore it has to coordinate related institutions for the common goal of sustainable use and development of urban ecosystems and habitats. The objective of this initiative is to reconcile unsustainable urban development pressures with the sustainable use of the exceptional biodiversity value, aquatic habitats and water quality of the basin with a UBR approach. The ecosystem approach is a key tool here for fostering the sustainability of the aquatic habitat and biological diversity through management of the three functions of the BRs (see Chapter 3).

9.8.4 Expected results

The most significant impact of the UBR approach will facilitate the implementation of the protection status for the area with a global importance. This process will carry the advantages of overlapping objectives for water quality and sustainability of the aquatic habitat. An important direction of a UBR concept is the rehabilitation of degraded ecosystems, with the same importance as in the case of undisturbed ones. Therefore it will be influential for upgrading the watershed area, heathlands, wetlands, forests, streams running through the Omerli reservoir and surrounding areas. Another significant impact will be on groundwater resources, which are currently threatened by uncontrolled residential development and dispersed industrial facilities in the protection zones of the watershed. These developments are creating negative impacts by polluted surface runoff, uncontrolled water pumping and the mining of aquifers. Defined goals and controlled protection status with public participation, in other words, balanced interaction of nature, economy and socio-cultural environment, will have tangible benefits for ecosystem services. From the societal perspective, the UBR approach will have the potential to improve the economic status of disadvantaged groups, such like as low-income households, recently immigrated populations (less experience with urban life), the elderly and others, by taking part in eco-sensitive economic activities.

9.8.5 Project implementation and stakeholders

Comprehensive biodiversity analysis will be necessary in the core zone of the proposed UBRs. Management and development functions as well as BR zone definitions are summarized in Table 9.4. Major participants, who will take part in the decision-making

Table 9.4 Recommended participants, land uses and possible impacts of the proposed Omerli Watershed UBR

Participants	UBR zones/land uses	Impacts of UBR Functions
MGI, ISKI, UNESCO, MoEF, CC, DHKD, ITU and IU	Core: Absolute and Short Distance Conservation Zones of the watershed have to be used for only educational and scientific research purposes to protect distinctive biodiversity and water quality. Buffer: Medium and Long Distance Conservation Zones of the watershed can be developed as organic farming, eco-tourism related land uses, urban agriculture (community gardens for disadvantaged groups), biodiversity related theme parks. Transition: Neighbouring open spaces with high biodiversity value, degraded sub-ecosystems and habitats, and urban areas serviced by the watershed for drinking water supply can be developed for ecosystem sensitive urban land uses and urban open space developments, stream and habitat restoration projects, organic farming, urban agriculture (community gardens for the urban poor and the neighbouring communities).	Conservation: Having protection status for the globally endangered endemic species and nationally important habitats which are under threat of illegal developments as well as new housing estates. Development: Protecting water quality, creating new jobs, and open spaces for the well-being of disadvantaged groups (urban poor, migrants, etc.). Logistics: Environmental awareness programmes, knowledge-information sharing at regional, national and global levels.

Source: Excerpted from Tezer, 2006.

process collaboratively, are listed in the table. Buffer and transition zones may be assumed as the medium- and long-distance protection zones (which have to be reassessed regarding the habitat characteristics) of watershed regulation for the Omerli Watershed IPA/IBA.

The conservation function of UBRs has to be extended to degraded or potentially degraded areas in and out of core zones. Stream restoration efforts will have significant impact on the protection of biodiversity and water quality in the watershed area. MGI has a key role in the implementation of the UBR concept in the province. UNESCO's MAB Programme, local universities (ITU and IU), DHKD and other nature protection NGOs, Conservation Councils (CC) of Istanbul, Ministry of Environment and Forestry (MoEF)-Nature Protection Directory and also citizens' organizations should all take part in the process of developing and implementing UBRs for Istanbul.

A UBR can be part of a city, or surround a city, but it does not have to be designated as the whole city. Existing global environmental degradation calls for urgent attention by reconciling development pressures with ecological units and biodiversity in urban areas. Sustainable practices in buffers or transition zones of UBRs should be designed as tools for controlling mechanisms of unplanned development pressures as well as for degraded ecosystem rehabilitation. The method of fostering UBRs has to be compatible with sustainable urban development needs and should be focused on biodiversity conservation efforts (Tezer, 2005).

Designation processes in urban areas have to take into account primarily sustainable biodiversity conservation efforts in conjunction with the improvements of other sustainable development programmes. This approach will facilitate the definition to be applied to urban areas. Sustainable development programmes, such as Healthy Cities, Agenda 21, Sustainable Cities Programme, Ecological Cities and others, can be good connections with the UBR programme, but the focus has to be clear, as it is in the Statutory Framework, and biodiversity has to be the departing point. Additionally, existing BRs in and around urban areas may have an opportunity of developing their management programmes with a new perspective on UBRs. The most important feature of UBRs should be related to the restorative dimension of degraded ecosystems and biodiversity protection in urban areas. This approach has to be connected with local land-use development plans (Tezer, 2005). The network of natural communities in an urban area with a UBR approach has to be implemented on a regional scale with their urban-wide interaction, integration and planning purposes for local, regional and even global sustainability needs.

9.8.6 Conclusion and recommendations

Water quality is completely dependent on healthy ecosystems and sustainable land-use management in watersheds. The conservation of riparian and aquatic habitats of watersheds has to be the primary task to balance natural and socio-environmental interactions. The Omerli Watershed faces the most acute degradation of water quality and biological diversity with pressures of unsustainable urbanization, which have been taking place in and around of its drinking water sources. The Omerli Watershed has been proposed here as an Urban Biosphere Reserve (UBR) in an attempt to reconcile urban development, water quality and biodiversity conservation in a more sustainable way for being the most important watershed of Istanbul.

Designating a UBR will add tangible benefits to a city, such as improving biodiversity conservation, public participation, and benefits from ecological, economic, research and educational perspectives. UBRs will be influential in protecting biological diversity in and around urban areas for sustainable urban development. It is obvious that success in biodiversity conservation will add visible values to urban life through channels of supporting ecological services, public benefit, public awareness building, education purposes and economic gains.

Global programmes that advocate better development in environment, economy and society have been improving interactions and collaborations among these pillars. Water, biodiversity and poverty, viewed as alarming concerns of global reality, need the same attention as interdisciplinary collaborations in urban environments for global sustainability. UNESCO's MAB and IHP programmes have been promoting better management policies for water and biodiversity issues in urban areas. This collaboration and practices like UBR approaches will strengthen and improve the sustainability of urban aquatic habitats as being discussed here for the Omerli Watershed case.

9.9 DESCRIPTION OF THE ECOLOGY AND WATER MANAGEMENT IN THE PHOENIX METROPOLITAN AREA, ARIZONA, USA

Elisabeth K LARSON, Nancy B GRIMM, Patricia GOBER and Charles L REDMAN

School of Life Sciences
Global Institute for Sustainability,
Decision Center for a Desert City Arizona State University, PO Box 84601 Tempe AZ 85287 4601 USA

This case study describes current water management conditions and aquatic habitats in the Phoenix, Arizona, USA metropolitan area. Phoenix is a city of 3.8 million inhabitants and is growing rapidly in a hot, dry desert that has appropriated a substantial amount of surface water from the Salt, Verde, and Colorado rivers. Current threats to aquatic habitats include agriculture, hard engineering solutions, loss of riparian areas, groundwater pollution, subsidence, recharging effluent water, and greening the desert with unfamiliar artificial habitats to the ecosystem and urbanization. Currently, a lack of ecological information, complex legal and institutional structures and a lack of extensive community interest impede maintenance and protection of native desert aquatic habitats.

Figure 9.28 **An artificial lake in Phoenix**

9.9.1 Background

Environmental Context

Located in the northern Sonoran Desert of the southwestern USA, the Phoenix metropolitan area receives approximately 180 mm of precipitation a year, with an average January temperature of 12°C and an average July temperature of 34°C (Baker et al., 2004). Most rain is concentrated in two seasons: a summer 'monsoon' season with short, intense, localized thunderstorms and a winter rainy season characterized by frontal storms of longer duration and lower intensity. Given its hot, dry climate, the area experiences an average of 2 m of potential evapotranspiration annually. The city is situated in an alluvial valley surrounded by rugged mountain ranges typical of Basin and Range topography (Jacobs and Holway, 2004).

Historical Background

Despite its arid climate, humans have lived in the Phoenix valley since prehistoric times. The valley is situated at the base of more humid upland watersheds. Dryland rivers, the Salt and Verde, provided adequate surface water to support settlement in an area where precipitation falls short of evapotranspiration substantially. The complex civilization of the Hohokam, which was based on irrigated agriculture, persisted for more than 1,000 years (Fitzhugh and Richter, 2004). When Phoenix was re-established as a farming community in the late 1800s to support area mining and military outposts, the ruins of Hohokam canals were discovered, excavated, and expanded upon (Gober, 2006), creating vast areas of agricultural production, including citrus, dairy, alfalfa and cotton crops. Throughout the twentieth century, new tactics for stabilizing and increasing water supply to the valley included the establishment of large dams and an extensive canal network throughout the region (see Figure 9.28 and Figure 9.29). These structural solutions essentially eliminated in-stream flow of the region's rivers, except during extreme flood events.

General Characteristics of the Urban Area

Phoenix is one of the fastest growing cities in the US, increasing from approximately 300,000 in 1950 to greater than 3.7 million in more than 20 municipalities in 2004, with a population density of 1074 people/km^2. Models predict that in 2025, the population will be exceed 6 million, representing a 280% change since 1980 (Jacobs and Holway, 2004), and nearly all of the undisturbed and agricultural lands will have been developed to urban land-uses (Jenerette and Wu, 2001). The growing population strains the existing water infrastructure and supply. More than half of the water supply comes from surface waters. With few geographical barriers to expansion, growth has been largely in an outward direction, estimated at almost 0.8 km per year (Gober and Burns, 2002). Most new construction has been the result of conversion of agricultural to residential use, but increasingly, new areas of desert are being transformed into housing developments. This rapid expansion of the urban area has often resulted in the destruction of desert washes, but has also led to the creation of numerous artificial lakes (see Figure 9.28). Thus, the demand for water stems from both necessity and strong aesthetic and recreational desires.

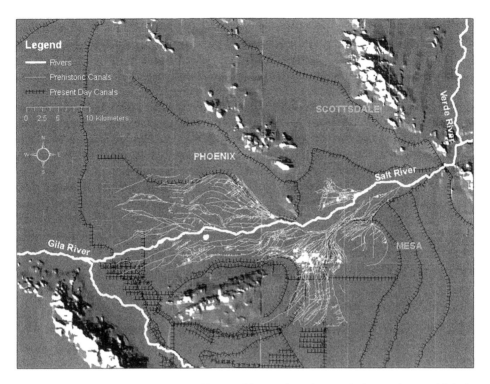

Figure 9.29 Major water features in the Phoenix Metropolitan Area, including prehistoric Hohokam Canals

Note: The Salt and Verde Rivers are tributaries to the Gila River, which in turn feeds into the Colorado River 100 s of km away.

Urban Water Management

Phoenix currently has a relatively diverse array of water resources available. The valley has access to $2.8 \times 10^9 \, m^3$ of water from the Salt and Verde rivers, groundwater, and the Colorado River. Sources comprise approximately 44% surface water (Salt and Verde Rivers), 39% groundwater, 12% Colorado River water, and 5% treated wastewater effluent (ADWR 1999). The Colorado River water is delivered by the Central Arizona Project canal (CAP canal), which moves water from the Colorado River eastward across the state some 450 km and uphill more than 700 m, to the cities of Phoenix and Tucson. The Phoenix valley has several hundreds of metres deep groundwater sub-basins that have been used to supplement surface water supply since the early 1900s. Declining water table levels have been occurring in some places since the 1940s, although early legislation (the 1948 Critical Area Groundwater Code) proved insufficient to slow the trend of increased well drilling. The state created the Groundwater Management Act (GMA) in 1980, a complex and ambitious regulatory plan to achieve safe-yield by 2025. As an Active Management Area (AMA), the Phoenix metropolitan area is subject to several management approaches (Jacobs and Holway, 2004). To date the GMA has had mixed success across the state. In the Phoenix AMA, the groundwater overdraft was reduced by approximately 40% between 1985 and 1995 (ADWR 1999), but 44 million m^3 per year are still overdrafted today (Baker et al., 2004). A major state programme now allows housing developers to purchase canal water for recharge into aquifers at a distance quite removed from their groundwater pumping, creating new ponds, which are continuously refilled to make up for infiltration and evaporation losses.

The above description of water sources to the Phoenix area is merely a broad overview; the intricacies are extremely complex and often opaque. Although the Arizona Department of Water Resources is charged with enforcing state regulations and has a general, regional perspective on the Phoenix AMA that includes surface water sources, its primary focus is on groundwater. There are more than 20 municipalities and agencies in the Phoenix metropolitan area in fact that make the practical management decisions; there are no standardized methods for accounting and no integrated management approaches that consider all water sources. With respect to surface water, aside from state and federal surface water quality standards (primarily focused on bacterial contamination), the main management effort in the Phoenix area has focused on stocking urban lakes with sport fishes. It is only recently that efforts have been made to consider maintaining or restoring native desert aquatic habitats.

9.9.2 Key aquatic habitat issues in urban water management

Very few locations throughout the urban area are typical desert aquatic habitats. Water, as a premium commodity, is moved miles to create desirable landscapes. What is desired is strongly influenced by the cultural backgrounds and experiences of stakeholders. Immigrants from more temperate climates, especially the US Midwest, represent a large proportion of Phoenix residents. With them, they bring memories of lush grasses, abundant vegetation, small ponds and lakes. Two aspects of city life reinforce the perception that such landscapes are sustainable. First, the regulation of water flows via damming and reservoirs has damped the strong pulse regime of desert hydrology, allowing available water to support vegetation year-round. Second, before the invention of air-conditioning, the cooling effect of increased evapotranspiration was an essential way to contend with extreme summertime temperatures. Thus, throughout Phoenix's history, new arrivals have seen a steady supply of water and abundant green growth, and many were attracted in the first place by promotions of the city as an idyllic place to retire, famous for its golf courses. There is no reason, given that such perceptions are actively encouraged, for newcomers to think of water as a limiting resource. Essentially, due to the creation of artificial aquatic habitats and lush gardens into the city landscape (see Figure 9.30), many residents no longer perceive or appreciate that they are living in the desert; for them, the desert exists only outside of the city (Farley-Metzger, personal communication; Gober, 2006).

Cultural preferences, along with the relatively easy access to a variety of water resources, have drastically changed the ecology of the Phoenix Valley. Demand for agriculture and later municipal uses have had a significant impact on contributing watersheds and downstream systems. Dams on all of the major rivers have eliminated pre-dam seasonal patterns of in-stream flow. Moreover, many flood mitigation efforts involved hard-engineering solutions including the lining of river channels. A significant amount of stormwater runoff is diverted to stormwater retention and detention basins associated with housing and commercial developments. These basins can serve

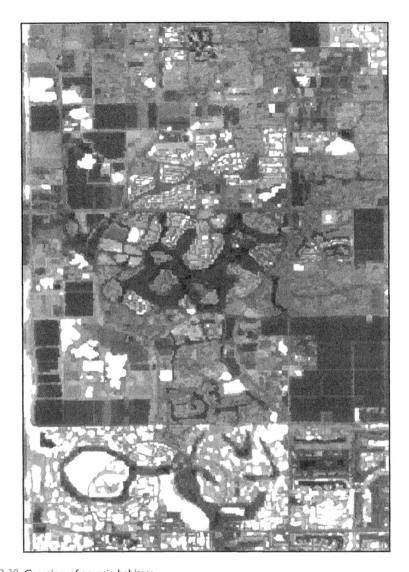

Figure 9.30 Creation of aquatic habitats

Source: ASTER satellite imagery of a suburban Phoenix area showing artificial lakes (in black).

several roles, providing flood mitigation, groundwater recharge, and recreational areas. The built environment has eliminated many natural flow paths (see Figure 9.31).

Historical modifications have resulted in an overall loss of riparian areas in some places, and a general shift in riparian community composition via bank stabilisation, the introduction of non-native plant species, and the decrease in total woody plant volume (Green and Baker, 2003). Some river and riparian habitats exist downstream of wastewater treatment plants, in river washes receiving stormwater runoff, and sites designated for groundwater recharge, but it is only recently that agencies have begun to

Figure 9.31 Photos of urban riparian areas in the Phoenix Metropolitan Area (See also colour plate 29)

Figure 9.31 (Continued)

consider ecological factors in management of aquatic systems in the Phoenix area. For example, the Rio Salado Project, funded by the City of Phoenix, the Flood Control District of Maricopa County, the Arizona Water Protection Fund, and the US Army Core of Engineers (US ACE) began the 'restoration' in 2001 of 240 ha of riverbed and riparian areas in central Phoenix. However, because the river flow regime has not been restored and groundwater levels have been lowered, there is not enough water naturally available to support these communities. Therefore, the project will include ground-water pumps, canals, and reservoirs to ensure adequate supply (City of Phoenix, 2005).

Meanwhile, further downstream at the 91st Avenue Wastewater Treatment Plant (WWTP), billions of litres of treated effluent are released into the Salt River annually. This nutrient-rich water supports an extensive riparian area, but little groundwater recharge is occurring because the area already has high groundwater levels. From the management perspective, this water is going to waste: a Bureau of Reclamation officer says, 'we just can't keep dumping it in the stream and letting it go downstream.' So an $80 million pro-ject is in the works to pipe the water northwest and uphill to the dry Agua Fria riverbed to facilitate groundwater recharge. Officials note that in addition to recharging the aquifer, the addition of this water will help restore native riparian habitat along the Agua Fria (Landers, 2004).

Thus, there inevitably are complex tradeoffs within the urban ecosystem between various water management and environmental objectives (Grimm et al., 2004). The culmination of historical decisions leads more and more frequently to the creation of 'designed ecosystems' to satisfy particular goals: water recharge, habitat restoration, aesthetics, and recreation.

The Phoenix metropolitan area now has greater than 650 artificial lakes (E.K. Larson, unpublished data). In Gilbert there is a Riparian Institute and water ranch: eighteen recharge ponds for treated effluent and constructed riparian habitat (the area is not an historical wash or river) designed to attract birds and other wildlife (Edwards, 2001).

9.9.3 Objectives of the case study

Are the water supplies and management practices for the Phoenix metropolitan area sustainable? The crux of this question is that the answer depends on the interaction of a multitude of uncertain ecological, economic, social, and cultural variables, and the very question of whether sustainability includes the maintenance, restoration, or protection of aquatic habitats is as yet unanswered, as many people, including water managers, consider allowing water to flow downstream (as opposed to captured for human use) a 'waste'. Assessment of these variables is ongoing, but is hampered by significant technological, organisational and informational difficulties.

With respect to renewable (non-groundwater) sources, new analyses continue to emerge. Some authors, such as Gammage (2003), argue that populations as high as 7 million will be sustainable as long as there is a corresponding decrease in agriculture (a water-intensive land-use). His view does not incorporate any climatic variability. Morehouse et al. (2002) conclude that the variety of water resources available to Phoenix provide more of a buffer to short-term drought conditions than Tucson, but 'even if agricultural demand were eliminated entirely, drought conditions would still force the AMA to rely on non-renewable supplies to meet 43% of its needs'. However, one of the most basic underlying assumptions about the flexibility of Phoenix water resources, that Colorado River water will provide when the Salt and Verde are experiencing drought and vice versa, was recently challenged by a joint University of Arizona/Salt River Project (SRP) report. The report used tree-ring analysis to reconstruct drought cycle synchrony between the two basins, and found that only two events in the 443 years analysed showed asynchronous flow (Hirschboeck and Meko, 2005), leading the manager of water resources at SRP to opine, 'our thought that the Colorado River would be able to bail us out is not a safe assumption anymore' (McKinnon, 2005).

Clearly, unlimited population growth rates are unsustainable due to accompanying environmental impacts and resource limitation. However, analysts predict the population of Phoenix to level off around 7 million (Gammage, 2003). Are there enough resources to support a population of this size without incurring serious environmental damage to groundwater and aquatic habitats, impairing resources for future generations? At the most basic level, research is still needed on human population trajectories, ecological impacts, climatic change, etc. There is a paucity of data on environmental outcomes of urban development, water use, landscaping practices, etc. at scales ranging from individual aquatic habitats to municipalities to watersheds to the Colorado River basin. As Gammage (2003) notes, 'because water's absence is the defining characteristic of a desert, its management becomes the defining activity of living in the desert'. Phoenix is not alone in addressing these questions; Fitzhugh and Richter (2004) estimate that '41% of the world's population lives in river basins where the per capita water supply is so low that disruptive shortages could occur frequently'. Evaluation of

Phoenix's sustainability, and implementation of steps to achieve it, will benefit not just Phoenix and the US Southwest but also rapidly growing cities throughout the world.

9.9.4 Potential stakeholders

From an academic and research perspective, there are numerous groups and institutions addressing water issues in the Phoenix Metropolitan area. The Central Arizona – Phoenix Long Term Ecological Research project (CAP LTER), a nationally funded programme now in its eighth year, seeks to expand and develop the necessary research tools and data to understand the long term, regional dynamics of the urban ecosystem (Grimm and Redman, 2004). Additional vital insight will be provided by the newly funded Decision Center for a Desert City (DCDC), a research institute at Arizona State University focused on establishing relationships between climatic conditions and water decision-making. The Arizona Water Institute, a joint collaboration with the University of Arizona and Northern Arizona University, focuses on water education, research, community assistance, and economic development in the State of Arizona as a whole. From a regulatory standpoint, the Arizona Department of Water Resources and the Arizona Department of Environmental Quality continue to assess Arizona water resources and ecosystems and implement regulatory plans and restoration activities. On a more local level, communities are beginning to work together to address management issues, as evidenced in the establishment of the East Valley Water Forum. However, integration of aquatic habitat concerns with water planning and management is still nascent.

9.9.5 Expected results

Collaboration of the stakeholders mentioned above will have multiple outcomes. Water and environmental agencies will gain a more comprehensive view of how water management and the ecology of aquatic ecosystems interact and influence one another. By taking a more integrative approach across disciplinary and institutional barriers, better planning and management decisions can be made that both provide sufficient water supply to the growing urban area while protecting, maintaining or restoring the unique aquatic habitats of the Sonoran Desert. The citizens, managers and institutions of the area must come to consensus on lifestyle issues (such as pool use and grassy landscaping), as well as environmental valuation of natural and manicured areas (lakes and golf courses) in order to build a unified vision of a sustainable Phoenix.

9.9.6 Conclusions

Beyond the traditional boundaries of basic natural science, urban ecological questions pose new challenges for researchers, as they necessitate interdisciplinary work between the natural and social sciences (Grimm et al., 2000). Anthropological and sociological questions about what makes Phoenix unique are intimately tied to the changing environmental setting. For instance, research has shown that plant diversity within the city is closely correlated with socio-economic factors such as family income and housing age (Hope et al., 2003). But socio-cultural values, like ecosystems, are mutable. For the Phoenix metropolitan area, water is perhaps the foremost integrator of these issues. Adequate assessment of regional sustainability and the means to achieve it require comprehending how values, economics, and the environment feedback to one another change over time.

REFERENCES

A.T.I. CSR. Presspali. 2000 Rilievi topografici e geognostoci del system arginale del fiume Adige, Autorità di Bacino del Fiume Adige, Studi. www.bacino-adige.it/studi_realizzati/25.

ADWR. 1999. Third Management Plan: 2000–2010 Phoenix Active Management Area, Arizona Department of Water Resources (ADWR), Phoenix, AZ.

Anonymous. 1999. *Istanbul Forever green!* (in Turkish), The Turkish Society for the Protection of Nature (DHKD), Istanbul, Turkey.

Aspinall, S. 1998. *A directory of the wetlands of the Middle East: United Arab Emirates.*

Autorità di Bacino del fiume Adige. 2006, *Il Territorio del bacino*, www.bacino-adige.it

Baker, L.A., Brazel, A.T. and Westerhoff, P. 2004. Environmental consequences of rapid urbanization in warm, arid lands: case study of Phoenix, Arizona (US). M. Marchettini, C. Brebbia, E. Tiezzi and L.C. Wadhwa (eds), *The Sustainable City III*. Advances in Architecture Series, MIT Press, Boston.

Bocian, J. 2004. Concept *of the biomass market development in the city of Lodz region* (trans. from Polish: Koncepcja organizacji rynku biomasy w rejonie m. Łodi). Expertise for the City of Lodz Office.

Bocian, J. and Zawilski, M. 2005. Ecohydrology concept – merging the ecology and hydrology for successful urban stream rehabilitation. Proceedings of the Urban River Rehabilitation Conference, Dresden, September 2005.

Braioni, M.G. and Ruffo, S. (eds). 1986. Ricerche sulla qualita delle acque dell'Adige , Mem. Museo Civ. St.Nat. Verona, No. 6, pp. 1–341.

Braioni, M.G., Bracco, F., Cisotto, P., Ghirelli, L., Villani, M.C., Braioni, A., Girelli, L., Masconale, M., Campeol, G. and Salmoiraghi, G. 2000. The biological – ecological and environmental landscape indices and procedures in the planning and sustainable management of the riverine areas: The case of the study of the river Dese and the river Adige. U. Maione, B. Maione Lehto and R. Monti (eds) *New Trends in Water and Environmental Engineering for Safety and Life.* Balkema, Rotterdam, pp. 97–110.

Braioni M.G. (ed.) 2001. Studies and Researches aimed to integration of the qualities of the river Adigeâ Final Report. National Basin Adige Authority, Biology Department of the Padova University, Evolutionary Department of the Bologna University, DAEST of the Architectural Institute of the Venice University, Trento and Bolzano Provinces, Veneto Region.

Braioni, M.G., De Franceschi, P., Braioni, A., Campeol, G., Caloi, S, Grandis, N., Pontiroli, A. and Ravanello, P. 2001a. New Environmental Indices for assessing bank quality in the restoration and the sustainable management of a river: the method. *Ecohydrology & Hydrobiology* 1 (1–2), pp. 133–54.

Braioni, M.G., Gumiero, B. and Salmoiraghi, G. 2001b. Leaf Bags and Natural Leaf Packs: Two Approaches to Evaluate River Functional Characteristics. *Internat. Rev. Hydrobiol.* 86 (4–5), pp. 439–51.

Braioni, M.G., Salmoiraghi, G., Bracco, F., Villani, M., Braioni, A. and Girelli, L. 2002a. Functional Evaluations in the Monitoring of the River Ecosystem Processes: the River Adige as a Case Study. *The Scientific World Journal*, vol.2, pp. 660–83, open access, search, research article. www.thescientificworld.com/SCIENTIFICWORLDJOURNAL/main/Home.asp

Braioni, A. and Braioni, M.G. 2002b. Fluvial Landscape evaluation: a method of analysis suitable to ecocompatible planning. G. Ceccu and U. Maione (eds). 2nd Conference New Trends in Water and Environmental Engineering for Safety and Life. (Capri 24–27 June 2002). Centro Studi Deflussi Urbani. D.I.I.A.R Politecnico di Milano. ISBN 88-900282-2-X: 1–14.

Braioni, M.G., Braioni, A. and Salmoiraghi, G. 2005a. Integrated Evaluation of the system 'River – fluvial corridor' by ecological and landscape – environmental Indices. The study cases: river Adige and Cordevole. Associazione Analisti Ambientali VQA n.2 – Studies, pp. 1–166

Braioni, M.G., Braioni, A. and Salmoiraghi, G. 2005b. A model and tools for the integrated management and planning of the system river – fluvial corridor. *Ecohydrology & Hydrobiology*, Vol., 5 (4), pp.124–38.

Braioni, M.G., Braioni, A. and Salmoiraghi, G. 2006. A model for the integrated management of river ecosystems. *Verh. Internat.Verein. Limnol.*, 29 (4), pp. 2115–23

Breil P., Lafont M., Namour, Ph., Perrin, J.F., Vivier, A., Bariac, B., Sebilo, M., Schmitt, L., Chocat, B, Aucour, A.M. and Zuddas, P.P. 2005. Dynamique du carbone et de l'azote en rivière dans un gradient rural-urbain. Ecco Sphère Colloque. Toulouse 5–7 December 2005. Article de conférence.

Breil, P, Grimm, N.B. and Vervier, Ph. 2007. Surface water-groundwater exchange processes and fluvial ecosystem function: an analysis of temporal and spatial scale dependency. P.J. Wood, D.M. Hannah and J.P. Sadler (eds) *Hydroecology and Ecohydrology: Past, Present and Future*.

CCA Environmental. 1999a. Final Scoping Report for the Moddergat River Improvement Scheme. Report prepared for Geustyn Loubser Streicher on behalf of Helderberg Municipality.

CCA Environmental. 1999b. Moddergat River Improvement Scheme. Environmental Management Plan (Construction Phase). Report prepared for Geustyn Loubser Streicher on behalf of Helderberg Municipality.

CIS FWD. 2003. Guidance on the best Practices in River Basin Planning, Working Project 2.9 WP2 4.3, pp. 1–88.

City of Phoenix. 2005. Rio Salado Habitat Restoration Directive 2000/60/EC of the European Parliament and of the Council of 23 October 2000 establishing a framework for Community action in the field of water policy.

D. Lgs. 1989. Legge 18 maggio 1989, n.183. Norme per il riassetto organizzativo e funzionale della difesa del suolo, *Suppl. Gazzetta Ufficiale* Serie gen. No. 120 del 28 May 1989.

D. Lgs. 1999. Decreto legislativo 11 maggio n. 152. Disposizioni sulla tutela delle acque dall'inquinamento e recepimento della Direttiva 91/271/CEE corcenente il trattamento delle acque reflue urbane e della Direttiva 91/676/CEE relativa alla protezione delle acque dall'inquinamento provocato dai nitrati provenienti da Fonti agricole, *Gazzetta Ufficiale* 29 May, No. 124.

D.Lgs. 2004. Decreto Legislativo Codice dei beni culturali e del paesaggio ai sensi dell'articolo 10 della legge 6 luglio 2002, No. 137, *Gazzetta Ufficiale* 24 February 2004, No. 45.

Dal Pra, A., De Rossi, P., Furlan, F., Siliotti, A. and Zangheri, P. 1991. Il regime delle acque sotterranee nell'alta pianura veronese, *Mem. Sc. Geologiche*, Vol. 42, pp. 155–83 Padova.

Davies, B.R. and Luger, M. 1994. Reviving our rivers. *Earthyear* 7, pp. 32–35.

Day, E. 2006. Al Wasit Nature Reserve (Ramtha Lagoon), Sharjah, United Arab Emirates. Follow-on assessment of Phase 1 wetlands and initial assessment of Phase 2 wetlands with recommendations for their rehabilitation and sustainable management. Freshwater Consulting Group report to Gary Bartsch International.

Day, E. and Ractliffe, G. 2002. Assessment of river and wetland engineering and rehabilitation activities within the City of Cape Town. Realisation of project goals and their ecological implications. Volume 1: Assessment process and major outcomes. Report to Catchment Management. Cape Metropolitan Council. City of Cape Town.

Day, E. and Ewart-Smith, J. 2005. Al Wasit Nature Reserve (Ramtha Lagoon). Sharjah, United Arab Emirates. Baseline study of surface aquatic ecosystems, with recommendations for their management and rehabilitation. Freshwater Consulting Group Report to Gary Bartsch International.

Day, E., Ractliffe, G. and Wood, J. 2005. An audit of the ecological implications of remediation, management and conservation or urban aquatic habitats in Cape Town, South Africa, with reference to their social and ecological contexts.

De Soyza, A.G., Vistro, N.B. and Boer, B. 2002. Sustainable development of mangroves for coastal Sabkha environments in Abu Dhabi, UAE. H-G. Barth and B. Boer (eds). *Sabkha Ecosystems*, Kluver Academic Publishers, Netherlands, pp 341–46.

Directive 2001/77/EC of the European Parliament and of the Council of 27 September 2001 on the promotion of electricity produced from renewable energy sources in the internal electricity market

Dogsé, P. 2004. Toward Urban Biosphere Reserves. C. Alfsen-Norodom, B.D. Lane and M. Corry (eds), *Urban Biosphere and Society: Partnership of Cities*, Annals of the New York Academy of Sciences, Vol. 1023, pp.10–48.

Duzzin B. 1986. Aspetti chimici – Caratteristiche fisiche e chimiche delle acque del fiume Adige e dei suoi affluenti in Provincia di Verona. M.G. Braioni and S. Ruffo (eds) *Ricerche sulla qualità delle acque dell'Adige* Mem. Museo Civ. St.Nat. Verona, N. 6, pp. 75–96.

Edwards, L. 2001. Gilbert's Riparian Preserves, Arizona Planning.

Emmons and Oliver Resources Incorporated, 2001. Benefits of wetland buffers: A study of functions, values and size. Prepared for the Minnehaha Creek Watershed District. Minnesota.

Fatta, D., Naoum, D. and Loizidou, M. 2002. Integrated environmental monitoring and simulation system for use as a management decision support tool in urban areas. *J. Environ. Management*, 64, pp. 333–43.

Fitzhugh, T.W. and Richter, B.D. 2004. Quenching urban thirst: Growing cities and their impacts on freshwater ecosystems. *Bioscience*, 54(8), pp. 741–54.

Gammage, G. 2003. Phoenix in Perspective: Reflections on Developing the Desert. Heberger Center for Design Excellence, Tempe, AZ.

GEOSS. 2006. Hydrogeological assessment of the extended Wasit Nature Reserve, Sharjah, UAE. Report G2006/08/01. Project No 205/11-71. Report to GBI.

Gnouma, R. 2006. Aide à la calibration d'un modèle hydrologique distribué au moyen d'une analyse des processus hydrologiques: application au bassin versant de l'Yzeron. Thèse de docteur de l'INSA-Lyon. 412p. http://csidoc.insa-lyon.fr/these/index.php?rub=0202.

Gober, P. 2006. Metropolitan Phoenix: Place Making and Community Building in the Desert. University of Pennsylvania Press, Philadelphia.

Gober, P. and Burns, E.K. 2002. The size and shape of Phoenix's urban fringe. *Journal of Planning Education and Research*, 21(4), pp. 379–90.

Grapentine, L., Rochfort, Q. and Marsalek, J. 2004. Benthic responses to wet-weather discharges in urban streams in southern Ontario. *Water Qual. Res. J. Canada* 39(4), pp. 374–91.

Green, D.A. and Baker, M.G. 2003. Urbanization impacts on habitat and bird communities in a Sonoran desert ecosystem. *Landscape and Urban Planning*, 63(4), pp. 225–39.

Grimm, N.B., Grove, J. M., Pickett, S.T.A. and Redman, C.L. 2000. Intregrated approaches to long-term studies of urban ecological systems. *BioScience* 50, pp. 571–84.

Grimm, N.B. and Redman, C.L. 2004. Approaches to the study of urban ecosystems: the case of central Arizona-Phoenix. *Urban Ecosystems* 7, pp. 199–213

Grimm, N.B. et al., 2004. Effects of urbanization on nutrient biogeochemistry of arid land streams. Ecosystems and Land Use Change, *Geophysical Monograph Series* 153, pp. 129–46.

Hancock, P.J. 2002. Human impacts on the stream-groundwater exchange zone. *Environmental Management* 29, pp. 763–81.

Hancock, P.J. and Boulton, A. J. 2005. Aquifers and hyporheic zones: Towards an ecological understanding of groundwater. *Hydrogeol. J.* 13, pp. 98–111.

Hancock, P.J., Boulton A.J. and Humphreys, W.F. 2005. Aquifers and hyporheic zones: Towards an ecological understanding of groundwater. *Hydrogeol. I.* 13, pp. 98–111.

Hirschboeck, K.K. and Meko, D.M. 2005. A Tree-Ring Based Assessment of Synchronous Extreme Streamflow Episodes in the Upper Colorado and Salt-Verde-Tonto River Basins, University of Arizona and Salt River Project, Tucson.

Holling, C.S. 1995. What barriers? What bridges? L.H. Gunderson, C.S. Holling, and S.S. Light (eds) *Barriers and Bridges to Renewal of Ecosystems and Institutions*. Columbia University Press, New York, pp. 3–34.

Hollis, G.E. 1975. The effect of urbanization on floods of different recurrence interval, *Water Resources Research*, 11, 3, pp. 431–35.

Hope, D., Gries, C., Zhu, W. Fagan, W.F., Redman, C.L., Grimm, N.B., Nelson, A.L., Martin, C. and Kinzig, A. 2003. Socioeconomics drive urban plant diversity. Proceedings of the National Academy of Science 100(15):8788–92. Online: http://www.pnas.org/cgi/doi/10.1073/pnas. 1537557100.

Hydrodata and Beta Studio. 1999. Studio degli acuiferi montani da Resia a Domegliara e degli acquiferi di pianura (CD-ROM) www.bacino-adige.it/pubblicazioni.asp.

Jacobs, K.L. and Holway, J.M. 2004. Managing for sustainability in an arid climate: lessons learned from 20 years of groundwater management in Arizona, USA. Hydrogeology Journal, 12(1), pp. 52–65.

Janauer, G.A. 1982. Die Gewässervegetation im 21. und 22. Wiener Gemeindebezirk (The Vegetation of water bodies in the 21. and 22. municipal districts of Vienna). ÖIR – Austrian Institute for Regional Planning (ed.) Bericht zum 2. Bearbeitungsschritt der Biotopkartierung Wien, pp.73–77.

Janauer, G.A. 2000. Ecohydrology: fusing concepts and scales. Ecological Engineering, Vol. 16, No. 9–16.

Janauer, G.A., Jungwirth, M. and Humpesch, U.H. (eds) 1985. Vorschläge für die ökologische Gestaltung des Stauraumes Wien. Danube Hydro Austria and Municipal Authorities Vienna MA 45 – Waterways Administration. Vols. 1–3.

Janauer, G.A. and Wychera, U. 2000. Biodiversity, succession and the functional role of macrophytes in the New Danube (Vienna, Austria). Arch. Hydrobiol. Supplement Large Rivers 12, pp. 61–74.

Jenerette, G.D. and Wu, J.G. 2001. Analysis and simulation of land-use change in the central Arizona-Phoenix region, USA. Landscape Ecology, 16(7), pp. 611–26.

Jones J.B. and Mulholland, P.J. 2000. Streams and ground waters. Academic Press, San Diego.

Kasahara, T. and Hill, A.R. 2006. Hyporheic exchange flows induced by constructed riffles and steps in lowland streams in southern Ontario, Canada. Hydrological Processes 20 (20), pp. 4287–305.

Klysik, K. and Fortuniak, K. 1999. Temporal and spatial characteristics of the urban heat island of Lodz, Poland. Atmospheric Environment 33, pp. 3885–95.

Lafont, M. 2001. A conceptual approach to the biomonitoring of freshwater: the Ecological Ambience System. Journal of Limnology, 60 (Suppl. 1), pp. 17–24.

Lafont M., Camus, J.C and Rosso, A. 1996. Superficial and hyporheic oligochaete communities as indicators of pollution and water exchange in the River Moselle, France. Hydrobiologia, 334, pp. 147–55.

Lafont, M. and Vivier, A. 2006a. Hyporheic zone, coarse surface sediments and oligochaete assemblages: importance for the understanding of the watercourse functioning: Hydrobiologia, 564, pp. 171–81.

Lafont, M., Vivier, A., Nogueira, S., Namour, Ph. and Breil, P. 2006b. Surface and hyporheic Oligochaete assemblages in a French suburban stream. Hydrobiologia, 564, pp. 183–93.

Lafont, M., Schmitt, L., Perrin, J.F., Breil P. and Namour, Ph. 2006c. Gestion de la Ressource en eau dan les hydrosystèmes périurbains. Rapport final de recherché. Cemagref ed.

Landers, J. 2004. Water resources – Arizona cities move forward with groundwater recharge plan. Civil Engineering, 74(1), pp. 20–21.

MAB Urban Group. 2003. Urban Biosphere Reserves in the Context of the Statutory Framework and the Seville Strategy for the World Network of Biosphere Reserves, Draft Policy Paper, June 2003, UNESCO Paris, http://www.unesco.org/mab/ecosyst/urban/doc.shtml, (last accessed: 26 July 2005).

McKinnon, S. 2005. UA Study: Dangers of drought heightened, The Arizona Republic, Phoenix, AZ.

Michlmayer, F. 1997. Flood control on the Danube, Vienna. Magistrat der Stadt Wien – MA 45.

Miliani, L. 1937. Le piene dei fiumi veneti e i provvedimenti di difesa. L'Adige. R. Acc.Naz. Lincei. Firenze, Vol. 7, No.1, pp. 1–303.

Morehouse, B.J., Carter, R.H. and Tschakert, P. 2002. Sensitivity of urban water resources in Phoenix, Tucson, and Sierra Vista, Arizona, to severe drought. *Climate Research*, 21(3), pp. 283–97.

Moriarty, P., Fonseca, C., Smits, S., Schouten, T., Butterworth, J. and Green, C. 2006. Learning Alliances for scaling up innovation and realizing integrated urban water management. Working paper (online) Available at www.switchurbanwater.eu.

Nicolis, E. 1898. Circolazione interna e scaturigini delle acque nel rilievo sedimentario – vulcanico della regione veronese e della finitica. *Accademia di Verona*, vol. 54, pp. 1–209.

Ozhatay, N., Byfield, A. and Atay, S. 2005. *122 Important Plant Areas of Turkey* (in Turkish), WWF Turkiye, Istanbul, Turkey.

Paour, F. and Hitier, P. 1998. Recomandation 40 on draft European Landscape Convention. Council of Europe, Congress of Local and Regional Authorities of Europe. V session (Strasbourg, 26–28 May 1998).

Pavoni, B., Duzzin, B. and Donazzolo, R. 1987. Contamination by chlorinated hydrocarbons (DDT, PCBs) in surface sediment and macrobenthos of the river Adige, *The Science of the Total Environment*, Elseviers Science Publishers B.V., Amsterdam, Vol. 65, pp. 21–39

Schmitt L. 2007. Hydro-geomorphological typologies of rivers as a basis for ecological monitoring and the management of aquatic ecosystems. (Submitted for publication in *Physics and Chemistry of the Earth*).

SIS. 2000. Population Censuses of the State Institute for Statistics, website: www.die.gov.tr, (last accessed: 24 July 2005).

Schlögel, G. 1997. Die Verbreitung und quantitative Erfassung der Gewässervegetation in der Lobau. Master Thesis, Vienna.

Statistical Yearbook for the Lodz Province. 2005. The Central Statistical Office.

Tezer, A. 1997. Modeling of Land-use and Transportation Interaction in the Case of Istanbul, (in Turkish), PhD Thesis, Institute of Science and Technology, Istanbul Technical University.

Tezer Kemer, A. 2004. Modeling of Land Use-Transportation Interaction in Istanbul, *AIZ ITU Journal of Faculty of Architecture*, Vol. 1, No. 2, pp. 12–25.

Tezer, A. 2005. The Urban Biosphere Reserve (UBR) concept for sustainable use and protection of urban aquatic habitats: case of the Omerli Watershed, Istanbul, *Ecohydrology and Hydrobiology*, Vol. 5, No. 4, pp. 311–22.

Tezer, A. 2006. Local developments versus global commitments: Reconciling impact of peripheral urbanization with natural environment, Istanbul case, IFoU 2006 Beijing International Conference, Modernization and Regionalism – Re-inventing the Urban Identity, Proceedings, Volume 1, pp. 322–32.

Tovey, N. and Aspinall, S. 2006. Wasit Nature Reserve. Sharjah. United Arab Emirates. Birds of Wasit Nature Reserve. Specialist report prepared for Gary Bartsch International.

Turri, E. and Ruffo, S. (eds) 1992. *Etsch Adige: the River, the People, the History*, Cierre Edizioni.

Schlögel, G. 1997. Die Verbreitung und quantitative Erfassung der Gewässervegetation in der Lobau. Master Thesis, Vienna.

UNCHS. 1996. *An Urbanizing World: Global Report on Human Settlements 1996*, United Nations Center for Human Settlements (HABITAT), Oxford University Press, Oxford, New York.

UN. 2004. *World Urbanization Prospects the 2003 Revision, Data Tables and Highlights*, United Nations, Department of Economic and Social Affairs, Population Division, New York.

Weyand, M. and Schitthelm, D. 2006. Good ecological status in a heavily urbanised river: is it feasible? *Water Science and Technology*, 53 (10), pp. 247–53.

Williams, W.D. 2002. Environmental threats to salt lakes and the likely status of inland saline ecosystems in 2025. *Environmental Conservation*. 29 (2), pp. 154–67.

Zalewski, M. (ed.) 2000. Ecohydrology. *Ecol. Eng.* Special issue 16, pp. 1–197.

Zalewski, M. 2006. Ecohydrology as a management tool. *Water and ecosystems: water resources management in diverse ecosystems and providing for human needs* 14–16 June 2005, Hamilton, Canada.

Zalewski, M. and Wagner, I. 2006. Ecohydrology – the use of water and ecosystem processes for healthy urban environments. Aquatic Habitats in Integrated Urban Water Management. *Ecohydrology and Hydrobiology.* Vol. 5. No 4, pp. 263–68.

Zalewski, M., Janauer, G.A. and Jolankai, G. 1997. Conceptual background. M. Zalewski, G.A. Janauer, G. Jolankai (eds) *Ecohydrology: A new paradigm for the sustainable use of aquatic resources.* International Hydrological Programme UNESCO, Paris, Technical Document in Hydrology 7.

Zawilski, M. 2001. Management of urban stormwater and storm overflow water with the use of small natural watercourses: a case study of Lodz. *Innovative technologies in urban drainage – NOVATECH 2001.* Lyon, 25–26 June 2001. Vol 2, pp. 699–706.

Index

Plate 1

Plate 2

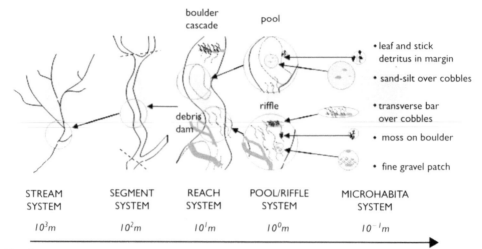

Plate 3

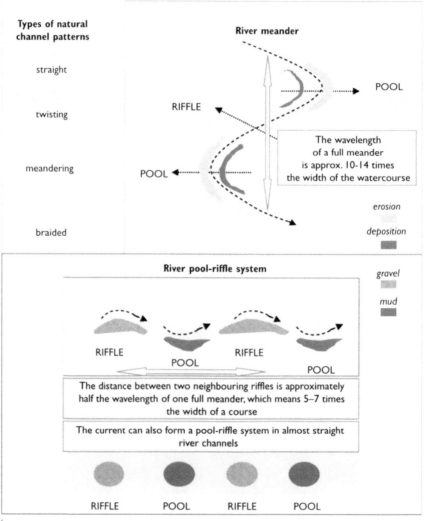

Plate 4

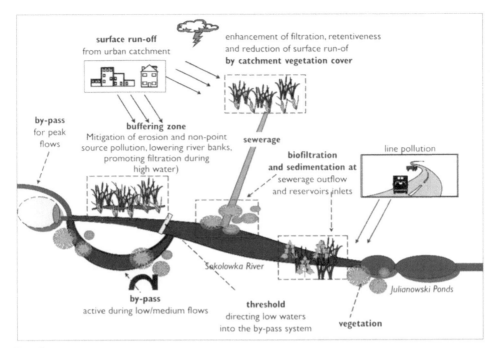

Plate 5

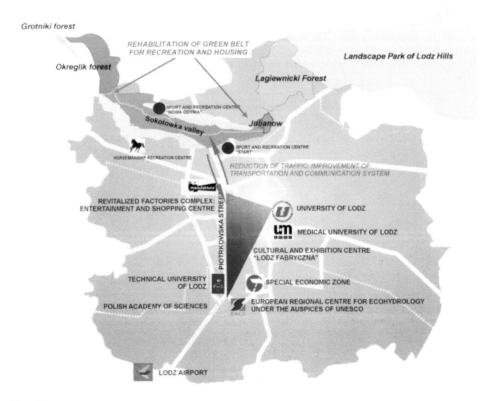

Plate 6

Plate 7

Plate 8

Plate 9

Plate 10

Plate 11

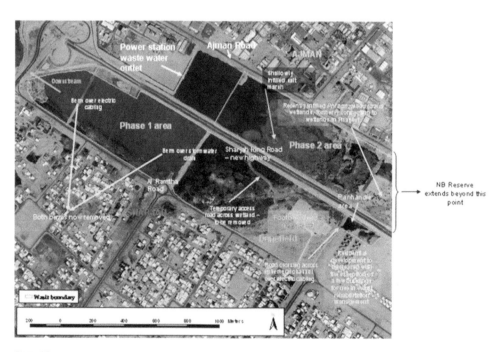

Plate 12

Plate 13

Plate 14

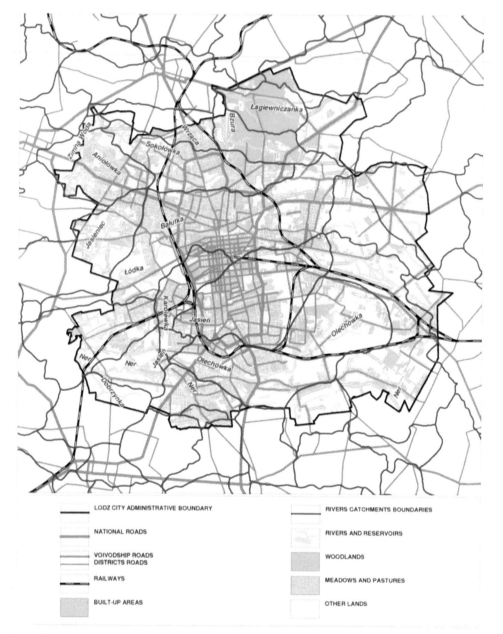

	LODZ CITY ADMINISTRATIVE BOUNDARY		RIVERS CATCHMENTS BOUNDARIES
	NATIONAL ROADS		RIVERS AND RESERVOIRS
	VOIVODSHIP ROADS		WOODLANDS
	DISTRICTS ROADS		
	RAILWAYS		MEADOWS AND PASTURES
	BUILT-UP AREAS		OTHER LANDS

Plate 15

Plate 16

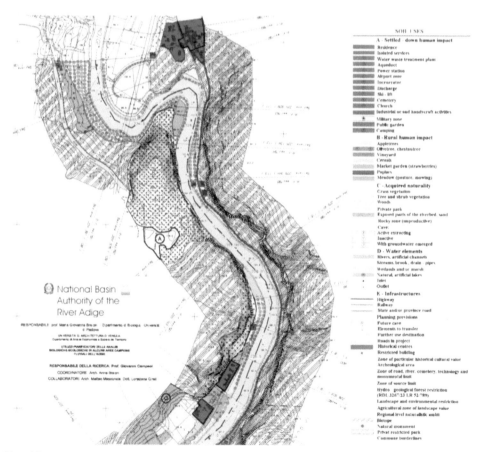

National Basin
Authority of the
River Adige

RESPONSABILE prof. Maria Giovanna Braioni   Dipartimento di Biologia, Universita
d Padova

UNIVERSITA DI ARCHITETTURA DI VENEZIA
Dipartimento di Analisi Economica e Scienze del Tempo

UTILIZZI PIANIFICATORI DELLA ANALISI
BIOLOGICHE-ECOLOGICHE IN ALCUNE AREE CAMPIONE
FLUVIALI DELL'ADIGE

RESPONSABILE DELLA RICERCA: Prof. Giovanni Campeol
COORDINATORE: Arch. Anna Braion
COLLABORATORI: Arch. Matteo Mascione, Dott. Loredana G.nel

Plate 17

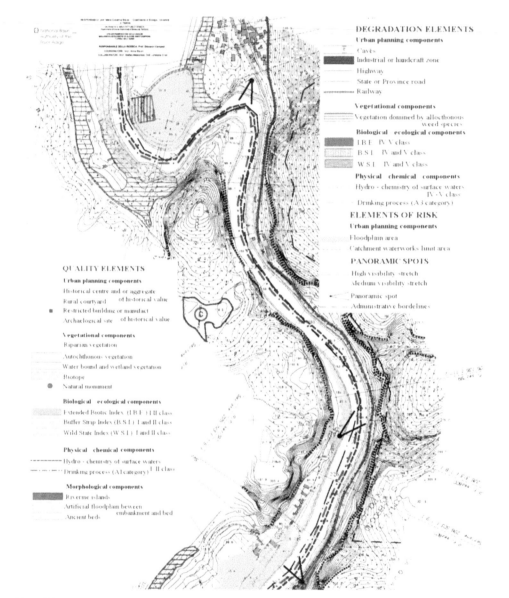

DEGRADATION ELEMENTS

Urban planning components
Caves
Industrial or handicraft zone
Highway
State or Province road
Railway

Vegetational components
Vegetation dominated by allochthonous weed species

Biological ecological components
I.B.E. IV-V class
B.S.I. IV and V class
W.S.I. IV and V class

Physical chemical components
Hydro - chemistry of surface waters IV - V class
Drinking process (A 3 category)

ELEMENTS OF RISK

Urban planning components
Floodplain area
Catchment waterworks limit area

PANORAMIC SPOTS
High visibility stretch
Medium visibility stretch
Panoramic spot
Administrative bordelines

QUALITY ELEMENTS

Urban planning components
Historical centre and or aggregate of historical value
Rural courtyard
Restricted building or manufact of historical value
Archaelogical site

Vegetational components
Riparian vegetation
Autochthonous vegetation
Water bound and wetland vegetation
Biotope
Natural monument

Biological ecological components
Extended Biotic Index (I.B.E.) I-II class
Buffer Strip Index (B.S.I.) I and II class
Wild State Index (W.S.I.) I and II class

Physical chemical components
Hydro - chemistry of surface waters
Drinking process (A1 category) I-II class

Morphological components
Riverine islands
Artificial floodplain beween embankment and bed
Ancient beds

Plate 18

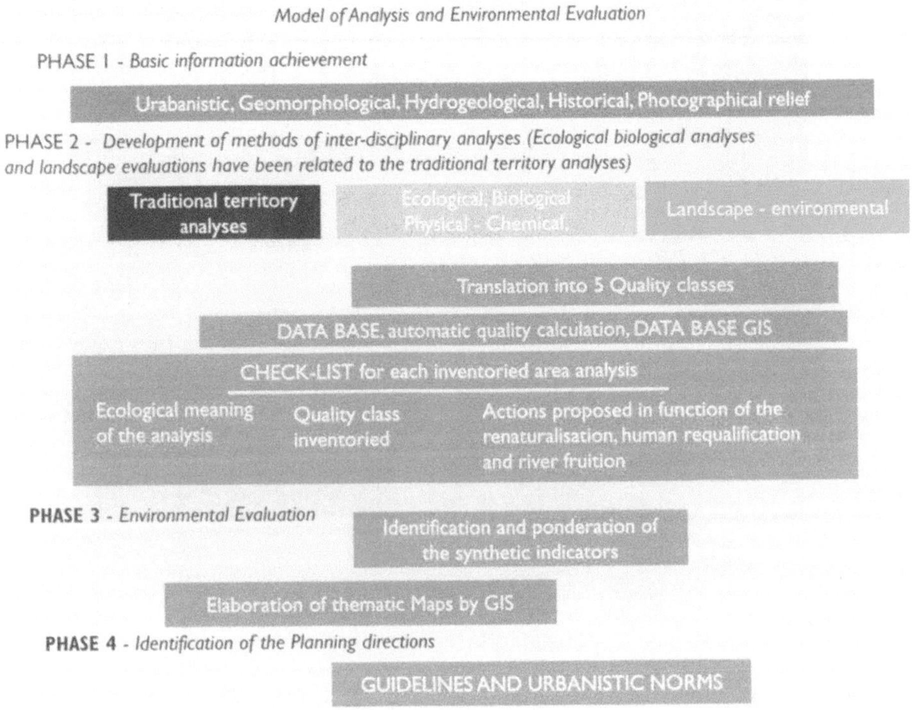

Model of Analysis and Environmental Evaluation

PHASE I - *Basic information achievement*

Urabanistic, Geomorphological, Hydrogeological, Historical, Photographical relief

PHASE 2 - *Development of methods of inter-disciplinary analyses (Ecological biological analyses and landscape evaluations have been related to the traditional territory analyses)*

Traditional territory analyses	Ecological, Biological Physical - Chemical,	Landscape - environmental

Translation into 5 Quality classes

DATA BASE, automatic quality calculation, DATA BASE GIS

CHECK-LIST for each inventoried area analysis

Ecological meaning of the analysis	Quality class inventoried	Actions proposed in function of the renaturalisation, human requalification and river fruition

PHASE 3 - *Environmental Evaluation*

Identification and ponderation of the synthetic indicators

Elaboration of thematic Maps by GIS

PHASE 4 - *Identification of the Planning directions*

GUIDELINES AND URBANISTIC NORMS

Plate 19

Plate 20

Plate 21

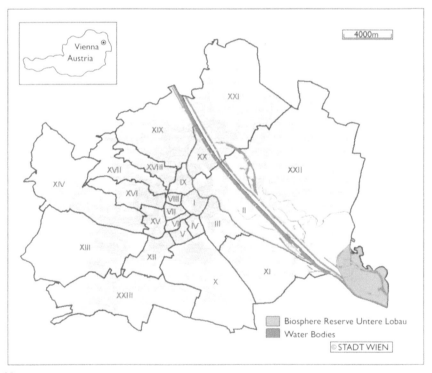

Plate 22

Plate 23

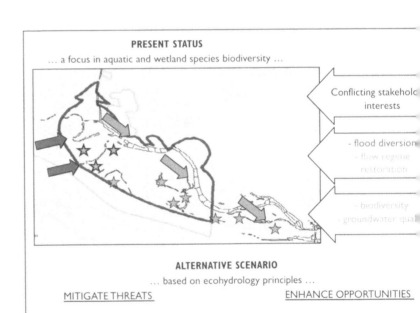

PRESENT STATUS

... a focus in aquatic and wetland species biodiversity ...

Conflicting stakeholc
interests

- flood diversion
- flow regime
restoration

- biodiversity
- groundwater qual

ALTERNATIVE SCENARIO

... based on ecohydrology principles ...

MITIGATE THREATS   ENHANCE OPPORTUNITIES

... increase dynamics / create new refuge habitats ...
... preserving ecosystem services under new hydrological regime ...

Plate 24

T - #0510 - 071024 - C16 - 246/174/12 - PB - 9780415453516 - Gloss Lamination